AF360830

Refrigeration, Air Conditioning and Heat Pumps

Refrigeration, Air Conditioning and Heat Pumps

Energy and Environmental Issues

Editor

Fabio Polonara

MDPI • Basel • Beijing • Wuhan • Barcelona • Belgrade • Manchester • Tokyo • Cluj • Tianjin

Editor
Fabio Polonara
Dipartimento di Ingegneria
Industriale e Scienze Matematiche (DIISM),
Universita' Politecnica delle Marche
Italy

Editorial Office
MDPI
St. Alban-Anlage 66
4052 Basel, Switzerland

This is a reprint of articles from the Special Issue published online in the open access journal *Energies* (ISSN 1996-1073) (available at: https://www.mdpi.com/journal/energies/special_issues/ Refrigeration_Air_Conditioning_Heat_Pumps_Energy_Environmental_Issues).

For citation purposes, cite each article independently as indicated on the article page online and as indicated below:

LastName, A.A.; LastName, B.B.; LastName, C.C. Article Title. *Journal Name* **Year**, *Volume Number*, Page Range.

ISBN 978-3-03943-823-5 (Hbk)
ISBN 978-3-03943-824-2 (PDF)

Contents

About the Editor

Fabio Polonara is Professor of Thermal Sciences at Università Politecnica (UNIVPM) delle Marche in Ancona, Italy. His research activity focuses on topics relating to refrigeration technology, the thermophysical properties of refrigerants and biofuels, renewable energies (with emphasis on biofuels), and energy planning. He has been a scientific project manager for research units working in the context of the European Union's JOULE, FLAIR, IEE, and MarieCurie schemes. Since 2015, he has been a member of TEAP (Technical and Economic Assessment Panel) and co-chair of RTOC (Refrigeration Technical Options Committee), both of which help the UNEP (United Nations Environment) to implement the Montreal Protocol. His research activities are documented in more than 200 papers.

Preface to "Refrigeration, Air Conditioning and Heat Pumps"

Refrigeration, air conditioning, and heat pumps (RACHP) have an important impact on the final energy uses of many sectors of modern society, such as residential, commercial, industrial, transport, and automotive. Moreover, RACHP also have an important environmental impact due to the working fluids that deplete the stratospheric ozone layer, which are being phased out according to the Montreal Protocol (1989). Last, but not least, high global working potential (GWP), working fluids (directly), and energy consumption (indirectly) are responsible for a non-negligible quota of greenhouse gas (GHG) emissions in the atmosphere, thus impacting climate change. To cope with this aspect, the Kigali Amendment of the Montreal Protocol (2016) has begun a phase down procedure for HFCs to be completed by the mid-21st century. All of these issues will pose great challenges to the RACHP industry over the next few decades, such as the search for new working fluids, ability to substitute high GWP HFCs, the safety aspects associated with the mostly flammable alternatives to high GWP HFCs, the expected growth of air conditioning in developing countries, and the subsequent increase in GHG emissions. The common ground for all of these challenges is that the energy efficiency of components and systems has to increase in order to keep energy consumption and GHG emissions associated with RACHP under control.

Fabio Polonara
Editor

Article

Energetic and Exergetic Analysis of Low Global Warming Potential Refrigerants as Substitutes for R410A in Ground Source Heat Pumps

Sergio Bobbo [1,*], Laura Fedele [1], Marco Curcio [1,2], Anna Bet [1], Michele De Carli [2], Giuseppe Emmi [2], Fabio Poletto [3], Andrea Tarabotti [3], Dimitris Mendrinos [4], Giulia Mezzasalma [5] and Adriana Bernardi [6]

[1] Istituto per le Tecnologie della Costruzione, Consiglio Nazionale delle Ricerche, I-35127 Padova, Italy; laura.fedele@itc.cnr.it (L.F.); la.rufanzia@gmail.com (M.C.); anna.bet@itc.cnr.it (A.B.)
[2] Dipartimento di Ingegneria Industriale, Università degli Studi di Padova, I-35131 Padua, Italy; michele.decarli@unipd.it (M.D.C.); giuseppe.emmi@unipd.it (G.E.)
[3] Hi-Ref S.p.A., I-35020 Tribano, Italy; fabio.poletto@hiref.it (F.P.); andrea.tarabotti@hiref.it (A.T.)
[4] Geothermal Energy Department, Centre for Renewable Energy Sources and Saving, 19009 Pikermi, Greece; dmendrin@cres.gr
[5] RED Srl. Via le dell'Industria 58B, I-35127 Padova, Italy; giulia.mezzasalma@red-srl.com
[6] Istituto di Scienze dell'Atmosfera e del Clima, Consiglio Nazionale delle Ricerche, I-35127 Padova, Italy; adriana.bernardi@isac.cnr.it
* Correspondence: sergio.bobbo@itc.cnr.it

Received: 1 July 2019; Accepted: 7 September 2019; Published: 16 September 2019

Abstract: In the European Union (EU), buildings are responsible for about 40% of the total final energy consumption, and 36% of the European global CO_2 emissions. The European Commission released directives to push for the enhancement of the buildings energy performance and identified, beside the retrofit of the current building stock, Heating, Ventilation, and Air Conditioning (HVAC) systems as the other main way to increase renewable energy sharing and overall building energy efficiency. For this purpose, Ground Source Heat Pumps (GSHPs) represent one of the most interesting technologies to provide energy for heating, cooling, and domestic water production in residential applications, ensuring a significant reduction (e.g., up to 44% compared with air-source heat pumps) of energy consumption and the corresponding emissions. At present, GSHPs mainly employ the refrigerant R410A as the working fluid, which has a Global Warming Potential (GWP) of 2087. However, following the EU Regulation No. 517/2014 on fluorinated greenhouse gases, this high GWP refrigerant will have to be substituted for residential applications in the next years. Thus, to increase the sustainability of GSHPs, it is necessary to identify short time alternative fluids with lower GWP, before finding medium-long term solutions characterized by very low GWP. This is one of the tasks of the UE project "Most Easy, Efficient, and Low-Cost Geothermal Systems for Retrofitting Civil and Historical Buildings" (acronym GEO4CIVHIC). Here, a thorough thermodynamic analysis, based on both energy and exergy analysis, will be presented to perform a comparison between different fluids as substitutes for R410A, considered as the benchmark for GSHP applications. These fluids have been selected considering their lower flammability with respect to hydrocarbons (mainly R290), that is one of the main concerns for the companies. A parametric analysis has been performed, for a reversible GSHP cycle, at various heat source and sink conditions, with the aim to identify the fluid giving the best energetic performance and to evaluate the distribution of the irreversibilities along the cycle. Considering all these factors, R454B turned out to be the most suitable fluid to use in a ground source heat pump, working at given conditions. Special attention has been paid to the compression phase and the heat transfer in evaporator and condenser.

Keywords: ground source heat pumps; low GWP refrigerants; energy analysis; R410A; R32; R454B

---

1. Introduction

In 2018 almost 12 million heat pumps were installed across Europe, and a large number of these were installed in Italy and France [1]. It should be stressed that this number accounts for just over 10% of operating heating systems and that gas boilers still occupy the majority of the market. Yet such a market is not sustainable from an environmental point of view. In this instance, according to EU Regulations [2,3], heat pumps come forward as an increasingly important player because they represent one of the main solutions in the direction to use more renewable energy for heating and cooling [4]. However, the application of Ground Source installations in the built environment is not well developed [5]. Previous studies were mainly focused on Air Source Heat Pumps and on refrigerant alternatives.

In the last two decades, the market of refrigerants has been dominated by hydrofluorocarbons (HFCs), which represent the biggest share of fluorinated greenhouse gases. As HFCs have a relatively high GWP and thus contribute to global warming when released into the atmosphere, the recent EU Regulation 517/2014 imposes a strong reduction of their total quantity to be marketed in the next 15 years, down to 20% of their annual marketed volume, compared to the year 2014 [6]. For instance, as a first step, starting from 2015, HFCs with GWP > 150 have been banned in domestic refrigerators [6].

This led to high prices of HFCs, urging the industry to find economically and environmentally sustainable alternatives of low GWP [7].

According to the type of application, different working fluids can be used in refrigeration cycles and the selection of the most suitable depends on practical and commercial purposes. Within the heat pumps and chillers sector, HFCs that are being replaced are basically R134a and R410A [8,9]. With a GWP respectively of 1043 and 2088, they are the most common high GWP refrigerants used at present. Their replacement is being undertaken to limit and control their emissions that contribute to global warming and climate change. Considering the best combination of cooling capacity and coefficient of performance (COP) results, Mota-Babiloni et al. [10] analyzed different HFC/HFO mixtures and showed good results for N-13, XP-10, and ARM-42A when substituting R134a, L41, and DR-5 as good alternative refrigerants to R410A.

In order to replace R410A, pure R32 was introduced in domestic air conditioners as a short-term option during the last couple of years. Despite being an HFC, R32 is characterized by much lower GWP than R410A (675 instead of 2088) and by only $\frac{1}{4}$ of the refrigerant charge needed for the same heating or cooling power output. Mota-Babiloni et al. [11] investigated its use in air conditioning and heat pump systems, confirming its slightly higher performance than R410A in terms of cooling and heating modes.

In the context of high temperature HPs, using R717, R365mfc, R1234ze(E), and R1234ze(Z) [12] and R1233zd(E) and 1336mzz(Z) [13], Kondou and Koyama [12] and Mateu–Royo et al. [13] evaluated the theoretical performance of different cycle configurations with hydrofluoroolefines in order to demonstrate the potential use of high temperature HPs to recover waste heat and reduce the primary energy consumption.

One of the task of the European Project GEO4CIVHIC is to analyze the possible HFCs alternatives suitable for the applications in the context of ground source heat pumps with domestic heating and cooling purposes. In doing that, many criteria have to be met, such as suitable thermodynamic properties, low flammability and toxicity, and stability in the system [14]. Merely substituting R410A with a lower GWP is not enough, because if the lower GWP refrigerant does not yield good performance in the system, it can lead to increased energy consumption, and thus to greater indirect emissions [15].

Wu and Skye [16] presented a survey on GSHPs using CO_2, NH_3, water, and hydrocarbons and evaluated advantages and disadvantages of natural refrigerants in terms of thermodynamic characteristics, operation in vapor compression GSHPs, and also flammability and toxicity. The parameter study presented by Aisyah et al. [17] is the first one that correlates an exergy and energy analysis of a heat pump system using R1224yd to the effect on the environment.

However, research on proper refrigerant to replace R410A in GSHPs for building heating, under varied ground conditions, is sparse. Researches on GSHPs are mainly related to GSHP design guidance and heat exchanger simulations [18,19] or to system application effects and control strategies [20,21]. The most common type of ground heat exchanger is the Borehole Heat Exchanger (BHE), a vertical pipe loop reaching depths of 50–200 m [22,23].

In this paper, the results of an analysis comparing short-term alternatives characterized by intermediate GWP (<1000) are reported, with the aim to identify transition solutions to be applied by EU companies in the period necessary to identify suitable refrigerants with very limited GWP (<150), that is the final goal to get low-environmental impact HPs. The selected transition fluid will be used within the project in a prototype and tested in a demo site. What it is expected from this analysis is to establish the most suitable fluid to be checked in pilot facilities. Moreover, it is expected that the solution that this study provides would allow Europe to increase its competeveness and assert its leadership in the field of Ground Source systems.

2. Methodological Approach of the Study

The purpose of this study is to simulate and compare the thermodynamic behavior of present and moderate GWP refrigerants in a reversible GSHP. Model simulation, carried out using Matlab software [24], was applied to predict the performance of the system under certain working conditions, besides irreversibilities in each component. Thermodynamic properties of fluids were calculated through Refprop 10.0 Database [25].

2.1. Refrigerant Selection

In this work, R410A was taken as the reference refrigerant and its performance in thermodynamic cycles were compared with those of alternative refrigerants. For its replacement, substitutes have to obtain the best compromise between energy efficiency and volumetric refrigeration capacity. Considering these factors, R32 and R454B were chosen as the most promising potential substitutes for R410A [26]. Both can be considered as transition solutions characterized by intermediate-GWP. According to the ASHRAE refrigerant classification standard, they are classified as A2L: low toxic and mildly flammable with burning velocities less than 10 cm·s^{-1} [27]. Their basic characteristics are given in Table 1.

Table 1. Basic characteristics of the selected fluids according to Refprop 10.0.

Fluid	GWP	ASHRAE Safety Class [27]	Composition (wt %)	T_{crit} (K)	P_{crit} (Mpa)	T Glide (K)	NPB (K)
R410A	2088	A1	R32/R125 (50/50)	343.32	4.770	0.05	212.15
R32	675	A2L	R32 (100)	351.55	5.816	-	221.15
R454B	466	A2L	R32/R1234yf (68.9/31.1)	350.15	5.041	1.5	222.25

2.2. Assumptions for the Thermodynamic Cycle

A simple vapor compression refrigeration system was considered to simulate the heat pump in heating mode. The system with the main components (compressor, condenser, expansion valve, and evaporator) is schematically shown in Figure 1. Secondary fluids are water in both heat exchangers, when temperature is above 0 °C. In case of low temperature of the fluid in the ground loops, secondary fluid in the evaporator is a brine (mixture of water and propylene glycol at fixed concentration of 30%).

The main goal of this work is studying GSHPs for retrofitting civil and historical buildings, with different low and high temperature terminals for heating. Since they have to be suitable for all buildings, climates and ground conditions considered in the project, different operating temperatures for the user and for the ground source were set to evaluate the thermodynamic performance.

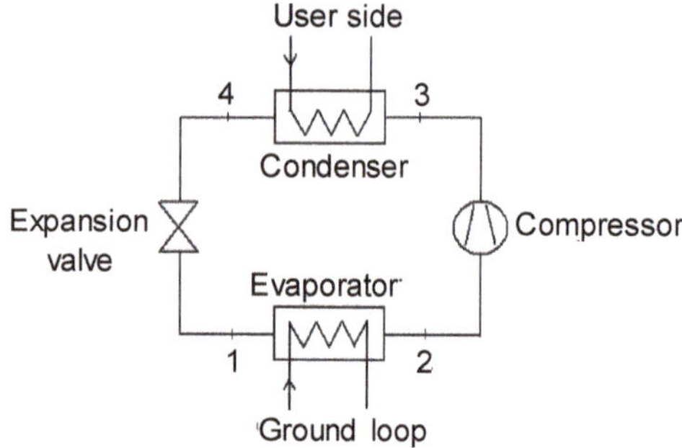

Figure 1. Layout of the heat pump system.

In relation to the ground loop, the temperatures of the heat carrier fluids can be variable depending on several factors like location, ground stratification, type of technology, and number and depth of the vertical ground source heat exchangers [28]. Furthermore, during the years of operation of the GSHP system, the temperatures could be very different if the boreholes field is not properly designed and the ground is affected by the so-called thermal drift [29].

Temperature differences of the secondary fluids through the heat exchangers were fixed, corresponding to typical values for heat pump already present in the market [30]. All the assumed system parameters and boundary conditions are specified in Section 2.6.

To model the heat exchangers (condenser and evaporator) a pinch analysis was applied. Pinch point position, i.e. the position in a heat exchanger where the temperature profiles of the fluids have the minimum temperature difference, is important to analyze heat transfer in thermodynamic cycle and it has to be determined accurately [31,32]. A very small minimum temperature difference between the temperature profiles of the fluids may cause an increase in costs, because much larger heat exchange surface areas are necessary. At the same time, bigger minimum temperature difference between the profiles increases exergetic losses in the heat transfer, thus decreasing the energy efficiency of the system. Pinch point position depends on the slope of the fluids temperature profile and on the superheating and subcooling assumed for the refrigerant. In the simulations, on the base of the assumptions made, the pinch point in the condenser is located at the inlet of the primary fluid; in the evaporator, it is located at the outlet of the primary fluid. These values are defined in Section 2.6.

In order to minimize irreversibilities associated with temperature difference, heat exchangers present a counter-current configuration of the fluids. This feature allows a better coupling between the temperature profiles and less exergetic losses. Considering the aims of the study, in this work, heat exchangers are considered as ideal, assuming a unit value for efficiency and no heat losses and pressure drops. An isenthalpic process was assumed for the expansion device, while a relatively complex procedure was followed to assume the isentropic efficiency of the compressor, as described in the next section.

Compressor

In order to accurately determine the performance of the thermodynamic cycle, isentropic efficiency of the compressor was carefully analyzed. As is well known, it represents the ratio between the work input to an isentropic process and the work input to the actual process, at the same inlet and exit pressures. Two approaches were considered to establish the influence of the isentropic efficiency on the cycle performance.

As a first step of the analysis, a constant value of the isentropic efficiency was assumed: this hypothesis ensures a fair comparison of the performances of different fluids, considering the same efficiency for all fluids and for all the working conditions.

The second step, more realistic, considered a variable isentropic efficiency, depending on the fluid, the selected commercial compressor and the working conditions. A scroll compressor (Bitzer GSU60120VL, Sindelfingen, Germany) available in the market and optimized for refrigerants R410A, R32, R454B, (and also R452B) was chosen, depending on the typical size of the GSHPs studied in the project. Working with this compressor at given conditions, each fluid has a different isentropic efficiency. The discharge temperatures were obtained through the Bitzer free online software 6.9 [33] and, indirectly, the isentropic efficiency was derived. This second approach helps to understand the different quality of the compression work and the heat losses occurring during the actual compression process, considering a real compressor present in the market.

Figure 2 shows the isentropic efficiency values for each fluid as a function of the user outlet temperature, i.e. the temperature of the secondary fluid at condenser outlet, once all the other boundary conditions are fixed. Assuming a constant value of the isentropic efficiency (Ref. line in Figure 2) for all fluids and all working conditions may lead to a not realistic evaluation of the compressor performance and therefore of the global performance of the thermodynamic cycle. Moreover, this assumption may cause a wrong selection of the most suitable refrigerant for a specific application with given working conditions.

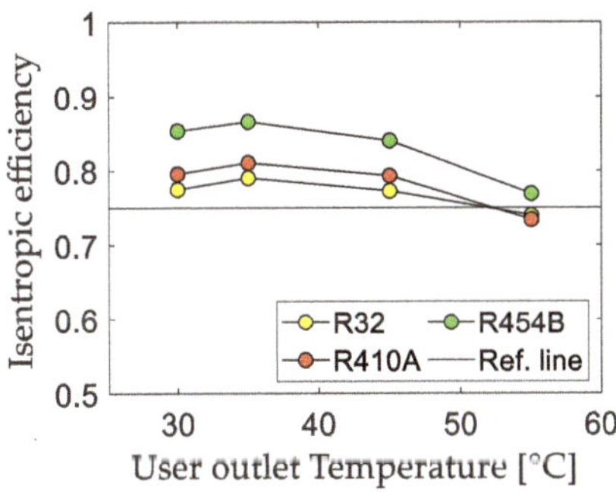

Figure 2. Constant value and actual values of isentropic efficiency at given boundary conditions calculated for the Bitzer GSU60120VL compressor.

2.3. Regenerative Cycle

With the aim to improve the performance of the base thermodynamic cycle, the addition of a liquid-line/suction-line heat exchanger (LLSL-HX) was evaluated. Thanks to the intra-cycle exchanger, the high-pressure refrigerant from the condenser is subcooled by the low pressure vapor entering the compressor [34]. This configuration is shown in Figure 3, with the description of the cycle in the P-h diagram.

As well known in the literature, high molecular mass fluids can take advantage from the regeneration because of their lower negative or their positive slope of vapor saturation curve in *T-s* diagram [35]. Subcooling of high-pressure liquid and superheating of low pressure vapor depend on the amount of heat transferred in the LLSL-HX. The maximum advantage is obtained considering a flooded evaporator, in which the refrigerant is not fully evaporated. Here, a two-phase mixture with 0.9 vapor quality has been considered as leaving low pressure fluid, where the vapor quality is defined from thermodynamics as the ratio between the vapor mass and the total mass of the mixture. The evaporation process is thus completed, together with the superheating of the vapor, in the LLSL-HX. The main benefit of this solution consists in a higher evaporation temperature and therefore in a reduction of the pressure ratio and of the compressor work.

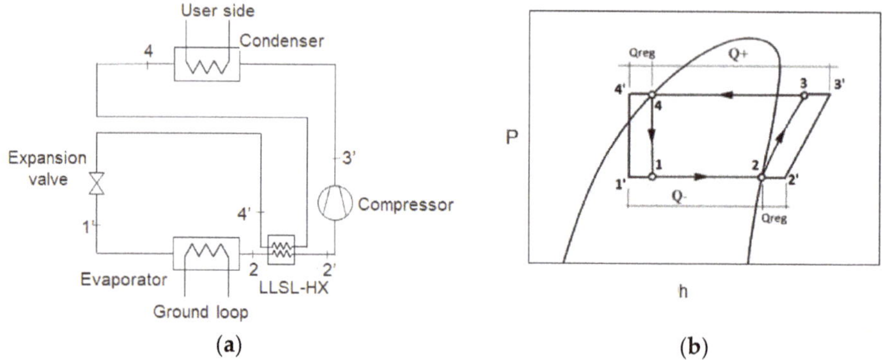

Figure 3. Compression cycle with (**a**) liquid-line/suction-line heat exchanger (LLSL-HX) and (**b**) the corresponding pressure-enthalpy diagram [36].

2.4. Energy Analysis

To evaluate the performance of the thermodynamic cycle, an energy analysis was carried out. Isentropic efficiency was calculated as follows (see Figure 2):

$$\eta_{is} = (h_{3is} - h_2)/(h_3 - h_2)$$

where h_{3is} and h_3 are the enthalpies at compressor discharge, respectively for isentropic and real compression, h_2 is the enthalpy at compressor suction.

Volumetric Heating Effect is another interesting parameter that represents the refrigerating effect per unit of swept volume. It provides information about the heat pump dimensions and about the required refrigerant charge.

$$VHE = \Delta h/v$$

where Δh is the enthalpy variation at the condenser and v is the refrigerant specific volume at compressor inlet.

The main energetic parameter used to compare the refrigerants efficiency is the coefficient of performance (COP) of the heat pump cycle, defined as the ratio between the heat supplied from the cycle to the hot reservoir ($\dot{Q}_{cond}$) and the required network input at the compressor ($\dot{W}_c$):

$$COP = \dot{Q}_{cond}/\dot{W}_c$$

where $\dot{Q}_{cond}$ represents the power absorbed by the working fluid at the condenser, exchanged with the user and set at 7000 W for every working condition.

The compressor power input $\dot{W}_c$ is calculated as follows:

$$\dot{W}_c = \dot{m}_{ref}(h_3 - h_2)$$

where $\dot{m}_{ref}$ is the refrigerant mass flow rate, which is known from the power exchanged at the condenser and the enthalpy difference at the condenser, h_3 is the enthalpy at compressor discharge and h_2 is the enthalpy at compressor suction.

2.5. Exergy Analysis

For a more comprehensive comparison of the refrigerant's performance, a detailed exergy analysis has been performed applying the general exergy theory described, e.g., in Reference [37]. This type of analysis allows to evaluate thermodynamic processes identifying the major sources of irreversibilities and then inefficiencies in energy exploitation. The optimization of a thermodynamic process has the purpose to minimize exergy losses, whereas energy, according to the first law of thermodynamics, cannot be destroyed and then no information on the quality of each thermodynamic process can be derived by energy balances. The overall system exergetic efficiency can be defined as follows:

$$\eta_{ex} = \dot{E}_{useful}/\dot{W}_c$$

where $\dot{E}_{useful}$ is the output exergy flux, i.e. the exergy absorbed by the user secondary fluid, and $\dot{W}_c$ is the input exergy flux to the system, supplied through the compression work.

The exergy absorbed by the user secondary fluid $\dot{E}_{useful}$ is obtained as:

$$\dot{E}_{useful} = \dot{m}_{user}(k_{u\_out} - k_{u\_in})$$

where $\dot{m}_{user}$ is the user flow rate of the secondary loop and k_{u_out} and k_{u_in} are the specific coenthalpies of the user fluid at the exit and at the entrance of the condenser. Coenthalpy is the potential of exergy flow and it is defined as:

$$k = h - T_a \cdot s$$

where Ta is a reference temperature, set at 5 °C and h and s are respectively the enthalpy and the entropy of the working fluid at the heat exchanger.

For each thermodynamic process in the cycle, exergy losses were calculated to evaluate their relative contribute to system energy efficiency. Exergy losses were calculated using the following equations:

- compressor $L_{comp} = \dot{W}_c - \dot{m}_{ref}(k_3 - k_2)$
- condenser $L_{cond} = \dot{m}_{ref}(k_3 - k_4) - \dot{m}_{user}(k_{u\_out} - k_{u\_in})$
- evaporator $L_{evap} = \dot{m}_{ground}(k_{g\_out} - k_{g\_in}) - \dot{m}_{ref}(k_1 - k_2)$
- throttling valve $L_{valve} = \dot{m}_{ref}(k_4 - k_1)$
- LLSL-HX $L_{LLSL\text{-}HX} = \dot{m}_{ref}[(k_4 - k_4') - (k_2' - k_2)]$

where $\dot{m}_g$ is the water flow rate of the ground loop, got from the balances, and k_{g_out} and k_{g_in} are the specific coenthalpies of the geothermal fluid at the exit and at the entrance of the evaporator.

2.6. Assumed System Parameters

The following assumptions have been made during the analysis:

- exchanged power at the condenser set at 7000 Watt.
- superheating at evaporator outlet should be as low as possible, but higher than zero to prevent liquid entering the compressor. Hence, a constant value of 6 K was assumed.
- subcooling in the condenser is another design parameter, fixed at 5 K.
- Pinch Point temperature difference $\Delta t_{pp} = 3\,°C$.

Simulations were run by varying both inlet and outlet temperature of the ground heat source (evaporator) and the user (condenser) secondary fluids (Table 2).

Table 2. Working conditions of the secondary fluids.

Conditions	Inlet Evaporator Temperature (°C)	Outlet Evaporator Temperature (°C)	Inlet Condenser Temperature (°C)	Outlet Condenser Temperature (°C)
1	0	−3	25	30
2	3	0	30	35
3	7	4	40	45
4	10	7	50	55

3. Results and Discussion

The main results of the analysis are summarized below and compare the performance of the thermodynamic cycle using R410A and those of the potential alternative fluids R32 and R454B. Diagrams are referred to the extreme working conditions for the secondary fluid circulating in the ground loop, i.e. inlet/outlet temperatures (T_g) at the evaporator:

- $T_g = 0/-3\,°C$
- $T_g = 10/7\,°C$

For these two sets of secondary fluid temperatures, performance of the heat pump producing domestic water at 4 different user secondary fluid outlet temperatures (from 30 to 55 °C) are shown.

3.1. Base Configuration

3.1.1. Isentropic Efficiency of the Compressor

Considering that isentropic efficiency is different for each fluid and working condition, as came out from previous analysis, it affects the performance of the thermodynamic cycle in different way. Figure 4 shows that compressor isentropic efficiency has a similar trend but, at the same time, different values for each fluid. R454B is the fluid that can guarantee the highest isentropic efficiency at all user temperatures.

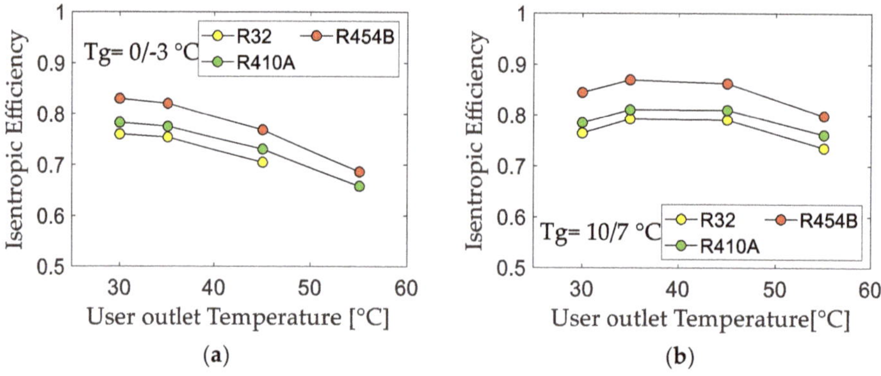

Figure 4. Isentropic Efficiency. (**a**) $T_g = 0/-3\,°C$; (**b**) $T_g = 10/7\,°C$

It is also interesting to notice how the shape of trends is different for different temperature levels of the heat carrier fluid of the ground loop (T_g). This variation depends on the nominal design limits of the compressor and is due to the fact that when compressor works with too high or too small pressure ratio, the efficiency of the compression process decreases.

3.1.2. Coefficient of Performance (COP)

Figure 5 shows that, for both extreme working temperature levels of the ground loop secondary fluid, COP decreases with the increasing of the user secondary fluid outlet temperature at the condenser, as expected due to the increase of pressure ratio. For the same reason, COP is lower when $T_g = 0/-3\,°C$ than when $T_g = 10/7\,°C$, considering all other conditions fixed.

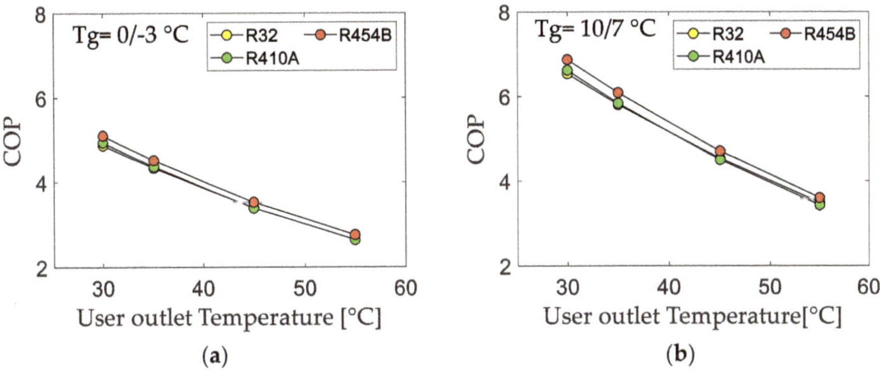

Figure 5. Coefficient of performance (COP). (**a**) $T_g = 0/-3\,^\circ$C; (**b**) $T_g = 10/7\,^\circ$C

R454B has slightly higher (around 5%) COP than R32 and R410A (similar each other) in all working conditions.

3.1.3. Volumetric Heating Effect (VHE)

The results of the VHE are summarized in Figure 6. As it can be seen, R32 has the highest VHE. Thus, it needs lower volumetric flow rate than the other fluids to exchange the same thermal power. R454B, vice versa, has the smallest value of the volumetric heating effect because it is a mixture of HFC R32 and HFO R1234yf having relatively low density. R410A has an intermediate behavior. It is interesting to note that the trend of VHE for each fluid is almost independent from the user secondary fluid outlet temperature at the condenser, that is from the pressure ratio. The trend of the volumetric heating effect is very important because it gives information about the required volumetric mass flow rate and therefore about the size of the heat pump and pipes.

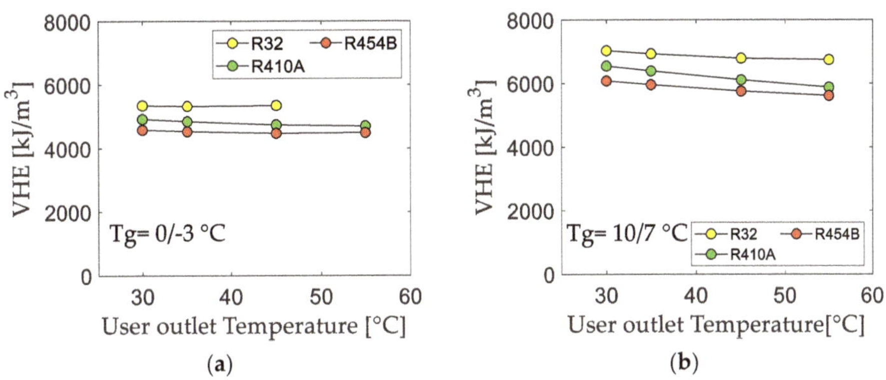

Figure 6. Volumetric heating effect. (**a**) $T_g = 0/-3\,^\circ$C; (**b**) $T_g = 10/7\,^\circ$C

3.1.4. Exergetic Efficiency

Exergetic efficiency trend is shown in Figure 7. As for the isentropic efficiency, the value is higher for higher temperatures of the ground loop fluid. R454B always stands as the refrigerant with the best performance, as from first law analysis. The trend of the performance is not linear because exergetic efficiency of components influences the heat transfer process differently according to the working temperatures.

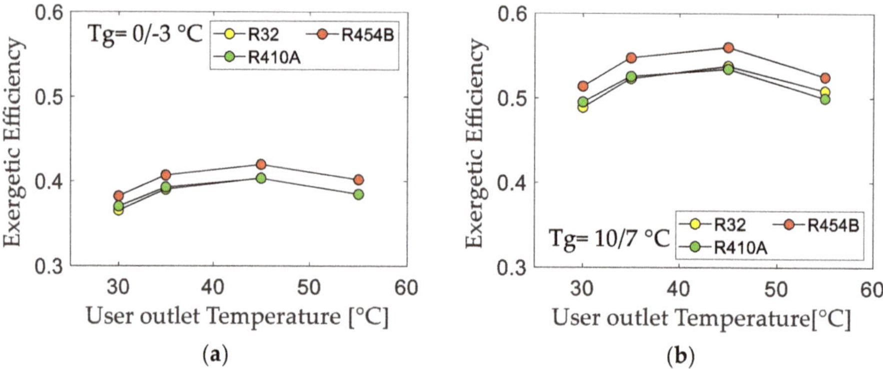

Figure 7. Exergetic Efficiency. (**a**) $T_g = 0/{-3}\,°C$; (**b**) $T_g = 10/7\,°C$.

3.1.5. Exergetic Losses

Exergetic analysis is useful because it allows to evaluate the distribution of exergetic losses and thus the inefficiencies of the heat pump components. Figure 8 shows what are the exergy losses generated in each component of the heat pump at a specific condenser condition for the lowest and the highest working temperature levels at the evaporator. In the case of $T_g = 0/{-3}\,°C$, it is evident that compression is the thermodynamic process with the highest exergy losses, followed by throttling. Even for the case with the highest evaporation temperatures ($T_g = 10/7\,°C$), and thus the lowest temperature difference between evaporation and condensation temperatures, compression and throttling are the worst processes, even if with lower differences from condensation and evaporation. Amongst the three refrigerants here considered, R454B showed total exergy losses lower than the other fluids: this result is mainly due to the much lower exergy losses in the compression process. In general, to improve the overall efficiency of the heat pump, efforts should be addressed to improve the efficiency of compression and throttling processes.

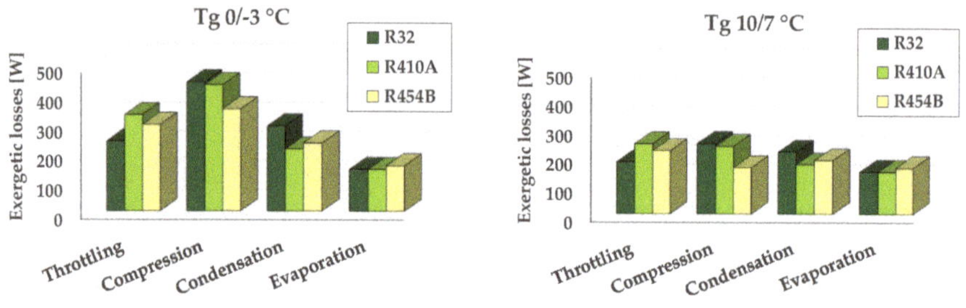

Figure 8. Exergy losses generated in the heat pump components considering inlet condenser temperature $40\,°C$ and outlet condenser temperature $45\,°C$. (**a**) $T_g = 0/{-3}\,°C$; (**b**) $T_g = 10/7\,°C$.

3.2. Regenerative Configuration

Regenerative configuration can increase the performance of the base cycle, depending on fluid and working conditions. In this configuration, a vapour quality of 0.9 was assumed for the working fluid at evaporator outlet. Then, the refrigerant moves to the LLSL heat exchanger, where it completes the evaporation and exits 5°C superheated.

The consequences at the evaporator are:

- pinch point moves at refrigerant inlet in the evaporator;

- evaporation temperature increases;
- temperature profiles are closer (Figure 9).

When the conditions are favorable, regeneration process allows to decrease exergetic losses and compression work. At the same time, it can improve the performance of thermodynamic cycle. In relation to isentropic efficiency, regeneration is not necessarily beneficial, because the reduction of the pressure ratio induced by the evaporating temperature increase is not always positive for the compression efficiency. However, compression efficiency worsening is less significant than compression work reduction and thus the overall effect can be positive.

Regenerative cycle performance is represented in terms of percentage deviation from the base cycle performance. Figures 10 and 11 highlight that regenerative configuration is beneficial for all fluids in every working condition in terms of both COP and exergetic efficiency. This means that using LLSL-HX improves energy performance in all cases and then is strongly suggested for these fluids and operative conditions. While R32 and R410A have similar improvement with respect to the base case, the increase of COP and exergetic efficiency is more evident for R454B with respect to the other fluids.

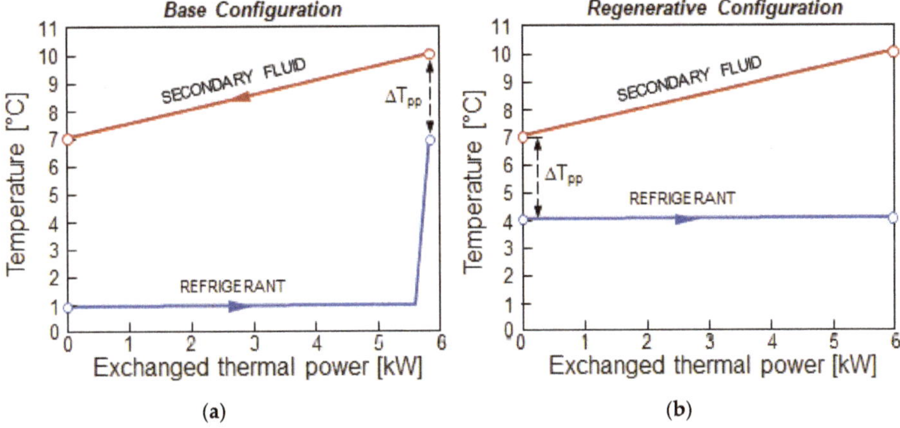

Figure 9. Temperature profiles in the evaporator for the two thermodynamic cycles. (a) Base Configuration; (b) Regenerative Configuration.

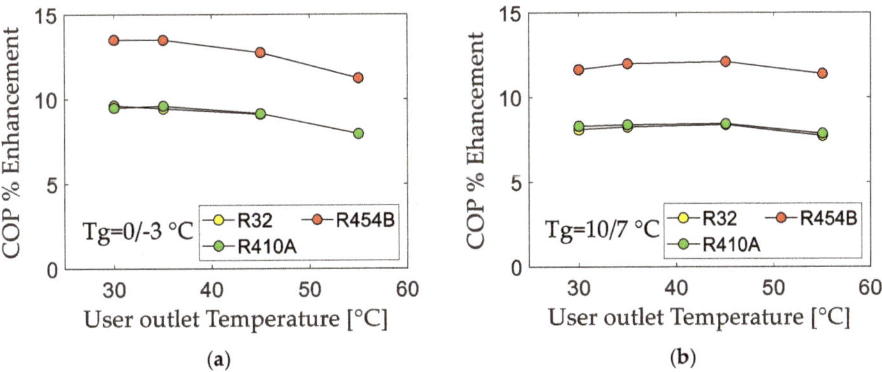

Figure 10. Percentage deviation of coefficient of performance (COP). (a) $T_g = 0/{-}3\,°C$; (b) $T_g = 10/7\,°C$.

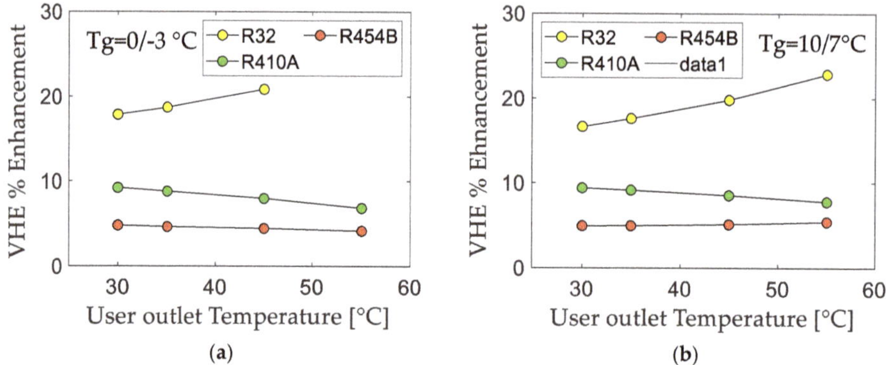

Figure 11. Percentage deviation of the Volumetric Heating Effect. (**a**) T_g = 0/−3 °C; (**b**) T_g = 10/7 °C.

The regenerative configuration leads also to increased VHE, as represented in Figure 11. In relation to VHE, R32 has the highest benefit with respect to the other refrigerants.

3.2.1. Exergetic Efficiency

As for the previous factors, exergetic efficiency benefits from the regenerative configuration as well. The increase of exergetic efficiency compared to the base cycle occurs for every fluid and every working temperature (Figure 12). Once again, R454B is the refrigerant that gets the most profit.

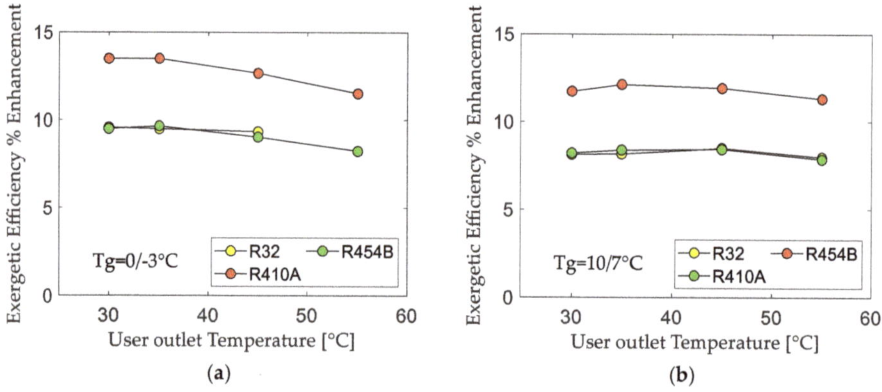

Figure 12. Percentage deviation of the exergetic efficiency. (**a**) T_g = 0/−3 °C; (**b**) T_g = 10/7 °C.

3.2.2. Exergetic Losses

The regenerative configuration allows to reduce some exergetic losses of the system, represented in Figure 13. Comparing Figures 8 and 13, a considerable reduction during the throttling and evaporation phase is noticeable.

During the evaporation process, losses decrease depending on the change of position of the pinch point. Since the pinch point moves at refrigerant inlet in the evaporator, the area between the two temperature profiles is greatly reduced, and the losses decrease (see Figure 9).

Throttling losses are reduced both because of the increase of evaporation temperature and because of the higher subcooling of liquid at the entrance of the valve.

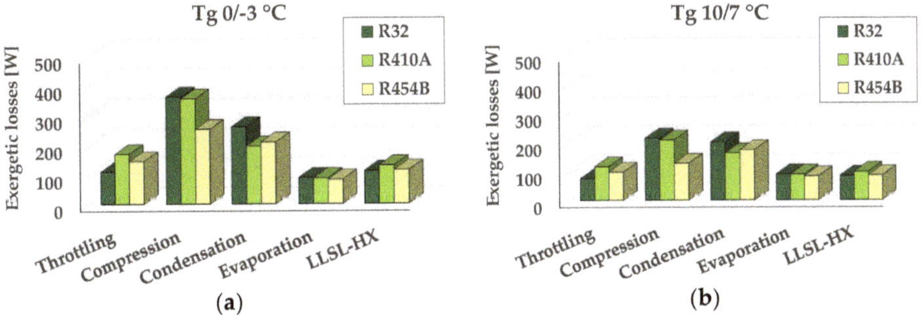

Figure 13. Exergy losses generated in the heat pump components, included liquid-line/suction-line heat exchanger (LLSL-HX), considering inlet condenser temperature 40 °C and outlet condenser temperature 45 °C. (**a**) T_g = 0/−3 °C; (**b**) T_g = 10/7 °C.

It also important to notice that the decrease varies according to the refrigerant and the complexity of the molecules.

Despite the fact that all the losses tend to decrease, it should be considered that in the regenerative configuration there is an additional exchanger, also characterized by exergetic losses. However, these losses do not give an important contribution to the total exergetic losses of the system.

4. Conclusions

As there are many factors influencing the performance of a thermodynamic cycle, it is necessary to carry out an overall analysis to identify the best refrigerant, amongst those with intermediate GWP (<1000), for the replacement of R410A in ground source heat pumps. Thus, a comprehensive analysis has been performed considering the first and the second laws of thermodynamics, as well as exergetic losses in the various components, compressor discharge temperature, and volumetric heating effect.

From the compressor point of view, the analysis has been performed considering both the case of fixed isentropic efficiency for all the considered refrigerants (R410A, R32 and R454B) and that of variable isentropic efficiency as calculated from a software reproducing the behavior of a commercial compressor working with all the considered refrigerants. It is interesting to note that the two approaches gave totally different results. In particular, in the more realistic case of variable isentropic efficiency, R454B showed to be the most promising refrigerant. Thus, assuming constant isentropic efficiency, even if in principle correct from the thermodynamic point of view, can be misleading with respect to the actual performance in a commercial compressor.

A comparison between base configuration and regenerative configuration of the cycle has also been performed, with the aim to evaluate the opportunity to install a LLSL-HX in the cycle. Significantly higher values of COP and exergetic efficiency than R410 and R32 are obtained with R454B in both cases, but with clear enhancements induced by the presence of the LLSL-HX with respect to the base configuration of the cycle.

Vice versa, R454B is characterized by a lower volumetric heating effect than R410A and R32. This can implicate a bigger size of components of the heat pump (e.g., compressor) and then higher costs. For sure, more studies and research have to be made in relation to R454B, which is basically unknown from the technical point of view, since the first compressors using this fluid have been produced in 2018. Moreover, R454B is a zeotropic mixture: therefore, it requests attention in heat exchangers optimization, since variations of the refrigerant temperature and composition can occur during the operation. Finally, in terms of flammability, it is classified as an A2L refrigerant: thus, the mass charged in the system should be limited, preventing use in large installations.

It is also important to underline that the analysis here performed is based on ideal assumptions (for example, effect of the pressure losses was not considered here) that could affect the results, if applied in real installations. The results found, however, provide a starting point for the selection of intermediate GWP replacement fluids.

Author Contributions: Conceptualization: S.B. and L.F.; Methodology: S.B., M.D.C. and M.C.; Software: M.C., A.T. and A.B. (Anna Bet); Validation: M.D.C., G.E. and D.M.; Formal Analysis: M.C. and A.B. (Anna Bet).; Investigation: M.C., A.B. (Anna Bet) and S.B.; Resources: F.P. and L.F.; Data Curation: G.M.; Writing-Original Draft Preparation: S.B., A.B. (Anna Bet) and M.C.; Writing-Review & Editing: S.B. and A.B. (Anna Bet); Visualization: S.B. and A.B. (Anna Bet); Supervision: M.D.C.; Project Administration: S.B. and L.F.; Funding Acquisition: A.B. (Adriana Bernardi).

Funding: This research was funded by Horizon 2020 (792355) and the APC was funded by ITC CNR.

Acknowledgments: The present study is realized within the project "Most Easy, Efficient and Low Cost Geothermal Systems for Retrofitting Civil and Historical Buildings" (Grant agreement ID: 792355) of the European Union's Horizon 2020 research and innovation program.

Conflicts of Interest: The authors declare no conflict of interest. The funders had no role in the design of the study; in the collection, analyses, or interpretation of data; in the writing of the manuscript, and in the decision to publish the results.

Nomenclature

COP	Coefficient of performance	-
GWP	Greenhouse Warming Potential	-
$\dot{E}_{useful}$	Exergy output	W
h	Enthalpy	J/kg
HFO	Hydrofluoroolefin	-
HP	Heat pump	-
HVAC	Heating, Ventilation and Air Conditioning	-
k	Coenthalpy	J/kg
L	Exergetic Loss	W
LLSL-HX	Liquid line/suction line heat exchanger	-
$\dot{m}_g$	Water flow rate of the ground loop	kg/s
$\dot{m}_{ref}$	Refrigerant mass flow rate	kg/s
$\dot{m}_{user}$	User mass flow rate	kg/s
s	Entropy	J/kg·K
T_a	Air temperature	°C
VHE	Volumetric heating effect	kJ/m^3
v	Specific volume	m^3/kg
$\dot{Q}_{cond}$	Power at the condenser	W
T_g	Water or brine temperature in the ground loop	°C
$\dot{W}_c$	Compressor power consumption	W
Greek Symbols		
η_{is}	Isentropic efficiency of the compressor	-
η_{ex}	Exergetic efficiency	-
Subscripts		
pp	Pinch point	

References

1. Nowak, T.; Westring, P. *European Heat Pump Market and Statistics Report*; The European Heat Pump Association AISBL (EHPA): Brussels, Belgium, 2018.
2. Council, E. Directive 2010/31/EU of the European Parliament and of the Council of 19 May 2010 on the energy performances of buildings. *Off. J. Eur. Union* **2010**, *153*, 13–35.
3. Directive (EU) 2018/844 of the European Parliament and of the Council of 30 May 2018 amending Directive 2010/31/EU on the energy performance of buildings and Directive 2012/27/EU on energy efficiency. Available online: https://eur-lex.europa.eu/legal-content/EN/TXT/PDF/?uri=CELEX:32018L0844&from=EN (accessed on 14 January 2019).

4. Bianco, V.; Scarpa, F.; Tagliafico, L. Estimation of primary energy savings by using heat pumps for heating purposes in the residential sector. *Appl. Therm. Eng.* **2017**, *114*, 938–947. [CrossRef]

5. Sanner, B.; Dumas, P.; Gavriliuc, R.; Zeghici, R. The use of geothermal energy for buildings refurbishment in Europe-technologies, success stories and perspectives. *Rev. Romana Inginerie Civila* **2013**, *4*, 183.

6. Directive (EU) 2014/517 of the European Parliament and of the Council of 16 April 2014 on fluorinated greenhouse gases and repealing Regulation (EC) 2006/842. Available online: https://eur-lex.europa.eu/legal-content/EN/TXT/PDF/?uri=CELEX:32014R0517&from=EN (accessed on 14 January 2019).

7. Mota-Babiloni, A. Analysis of low Global Warming Potential fluoride working fluids in vapour compression systems. Experimental evaluation of commercial refrigeration alternatives. Ph.D. Thesis, Jaume I University, Castellón de la Plana, Spain, 2016.

8. McLinden, M.O.; Kazakov, A.F.; Brown, J.S.; Domanski, P. A thermodynamic analysis of refrigerants: Possibilities and tradeoffs for Low-GWP refrigerants. *Int. J. Refrig.* **2014**, *38*, 80–92. [CrossRef]

9. Botticella, F.; De Rossi, F.; Mauro, A.W.; Vanoli, G.P.; Viscito, L. Multi-criteria (thermodynamic, economic and environmental) analysis of possible design options for residential heating split systems working with low GWP refrigerants. *Int. J. Refrig.* **2017**, *87*, 131–153. [CrossRef]

10. Mota-Babiloni, A.; Navarro-Esbrı, J.; Barragan-Cervera, A.; Moles, F.; Peris, B. Analysis based on EU Regulation No 517/2014 of new HFC/HFO mixtures as alternatives of high GWP refrigerants in refrigeration and HVAC systems. *Int. J. Refrig.* **2015**, *52*, 21–31. [CrossRef]

11. Mota-Babiloni, A.; Navarro-Esbrı, J.; Makhnatch, P.; Moles, F. Refrigerant R32 as lower GWP working fluid in residential air conditioning systems in Europe and the USA. *Renew. Sustain. Energy Rev.* **2017**, *80*, 1031–1042. [CrossRef]

12. Kondou, C.; Koyama, S. Thermodynamic assessment of high-temperature heat pumps using Low-GWP HFO refrigerants for heat recovery. *Int. J. Refrig.* **2015**, *53*, 126–141. [CrossRef]

13. Mateu-Royo, C.; Navarro-Esbrı, J.; Mota-Babiloni, A.; Amat-Albuixech, M.; Moles, F. Theoretical evaluation of different high-temperature heat pump configurations for low-grade waste heat recovery. *Int. J. Refrig.* **2018**, *90*, 229–237. [CrossRef]

14. Bobbo, S.; Di Nicola, G.; Zilio, C.; Brown, J.S.; Fedele, L. Low GWP halocarbon refrigerants: a review of thermophysical properties. *Int. J. Refrig.* **2018**, *90*, 181–201. [CrossRef]

15. Makhnatch, P.; Khodabandeha, R. The role of environmental metrics (GWP, TEWI, LCCP) in the selection of low GWP refrigerant. *Energy Procedia* **2014**, *61*, 2460–2463. [CrossRef]

16. Wu, W.; Skye, H. Progress in Ground Source Heat Pumps using natural refrigerants. *Int. J. Refrig.* **2018**, *92*, 70–85. [CrossRef] [PubMed]

17. Aisyah, N.; Alhmaid, M.I.; Nasruddin, S.; Lubis, A.; Saito, K. Parametric study and multi-objective optimization of vapor compression heat pump system by using environmental friendly refrigerant. *J. Adv. Res. Fluid Mech. Therm. Sci.* **2019**, *54*, 44–56.

18. Nam, Y.; Chae, H.B. Numerical simulation for the optimum design of ground source heat pump system using building foundation as horizontal heat exchanger. *Energy* **2014**, *73*, 933–942. [CrossRef]

19. Li, H.; Nagano, K.; Lai, Y.; Shibata, K.; Fujiimoto, H. Evaluating the performance of a large borehole ground source heat pump for greenhouses in northern Japan. *Energy* **2013**, *63*, 387–399. [CrossRef]

20. Hu, B.; Li, Y.; Mu, B.; Wang, S.; Seem, J.E.; Cao, F. Extremum seeking control for efficient operation of hybrid ground source heat pump system. *Renew. Energy* **2016**, *86*, 332–334. [CrossRef]

21. Emmi, G.; Zarrella, A.; De Carli, M.; Galagaro, A. An analysis of solar assisted ground source heat pump in cold climates. *Energy Convers. Manag.* **2015**, *106*, 660–675. [CrossRef]

22. Biglarian, H.; Abbaspour, M.; Hassan Saidi, M. Evaluation of a transient borehole heat exchanger model in dynamic simulation of a ground source heat pump system. *Energy* **2018**, *147*, 81–93. [CrossRef]

23. Luo, J.; Rohn, J.; Bayer, M.; Priess, A. Thermal Efficiency Comparison of Borehole Heat Exchangers with Different Drillhole Diameters. *Energies* **2013**, *6*, 4187–4206. [CrossRef]

24. The Mathworks Inc. *Matlab and Statistics Toolbox Release 2018a*; The Mathworks Inc.: Natick, MA, USA, 2018.

25. Lemmon, E.W.; Bell, I.H.; Huber, M.L.; McLinden, M.O. *NIST Standard Reference Database 23, Reference Fluid Thermodynamic and Transport Properties (REFPROP), version 10.0*; National Institute of Standards and Technology: Gaithersburg, MD, USA, 2018.

26. Pham, H.M.; Monnier, K. Interim and Long-Term Low-GWP—Refrigerant Solutions for Air Conditioning. In Proceedings of the International Refrigeration and Air Conditioning Conference at Purdue, West Lafayette, IN, USA, 11–14 July 2016.

27. ASHRAE. *ANSI/ASHRAE Standard 34-2016, Designation and Safety Classification of Refrigerants*; ASHRAE: Atlanta, GA, USA, 2016.

28. Sarbu, I.; Sebarchievici, C. Using Ground-Source Heat Pump Systems for Heating/Cooling of Buildings. In *Advances in Geothermal Energy*; InTechOpen: London, UK, 2016.

29. Capozza, A.; De Carli, M.; Zarrella, A. Design of borehole heat exchangers for ground-source heat pumps: A literature review, methodology comparison and analysis on the penalty temperature. *Energy Build.* **2012**, *55*, 369–379. [CrossRef]

30. Melinder, A. *Properties of Secondary Working Fluids for Indirect Systems (Secondary Refrigerants or Coolants, Heat Transfer Fluids)*; International Institute of Refrigeration: Paris, France, 2010.

31. Wu, W.; Zaho, L.; Ho, T. Experimental investigation on pinch points and maximum temperature differences in a horizontal tube-in-tube evaporator using zeotropic refrigerants. *Energy Convers. Manage.* **2012**, *56*, 22–31. [CrossRef]

32. Venkatarathnam, G.; Mokashi, G.; Murthy, S.S. Occurence of pinch points in condensers and evaporators for zeotropic refrigerant mixtures. *Int. J. Refrig.* **1996**, *19*, 361–368. [CrossRef]

33. Bitzer software v6.9.1.2074; Bitzer Kuhlmaschinenbau GmbH, 2013–2018. Available online: https://www.bitzer.de/websoftware/ (accessed on 4 February 2019).

34. Minh, N.Q.; Hewitt, N.J.; Eames, P.C. Improved Vapour Compression Refrigeration Cycles: Literature Review and their Applications to Heat Pumps. In Proceedings of the International Refrigeration and Air Conditioning Conference at Purdue, West Lafayette, IN, USA, 11–14 July 2016.

35. Babatunde, A.F.; Sunday, O.O. A Review of Working Fluids for Organic Rankine Cycle (ORC) Applications. *Mater. Sci. Eng.* **2018**, *413*, 012019. [CrossRef]

36. Domanski, P.A.; Didion, D.A.; Doyle, J.P. Evaluation of suction-line/liquid-line heat exchange in the refrigeration cycle. *Int. J. Refrig.* **1994**, *17*, 487–493. [CrossRef]

37. Borel, L.; Favrat, D. *Thermodynamique et Energetique: de L'energie a L'exergie, Vol. 1 and 2*; EPFL-PPUR: Lausanne, Switzerland, 2005.

Article

Energy Evaluation of Multiple Stage Commercial Refrigeration Architectures Adapted to F-Gas Regulation

Jesús Catalán-Gil *, Daniel Sánchez, Rodrigo Llopis, Laura Nebot-Andrés and Ramón Cabello

Department of Mechanical Engineering and Construction, Campus de Riu Sec., Jaume I University, E-12071 Castellón, Spain; sanchezd@uji.es (D.S.); rllopis@uji.es (R.L.); lnebot@uji.es (L.N.-A.); cabello@uji.es (R.C.)
* Correspondence: jcatalan@uji.es; Tel.: +34-964-72-8133

Received: 28 June 2018; Accepted: 18 July 2018; Published: 23 July 2018

Abstract: This work analyses different refrigeration architectures for commercial refrigeration providing service to medium and low temperature simultaneously: HFC/R744 cascade, R744 transcritical booster, R744 transcritical booster with parallel compression, R744 transcritical booster with gas ejectors, R513A cascade/R744 subcritical booster, and R513A cascade/R744 subcritical booster with parallel compression. The models were developed using compressor manufacturers' data and real restrictions of each system component. Limitations and operating range of each component and architecture were analysed for environment temperatures from 0 to 40 °C considering thermal loads and environment temperature profiles for warm climates. For booster systems, cascade with subcritical booster with parallel compression provide highest coefficient of performance (COP) for temperatures below 12 °C and above 30 °C with COP increases compared basic booster up to 60.6%, whereas for transcritical boosters, architecture with gas ejectors obtains the highest COP with COP increases compared to the basic booster up to 29.5%. In annual energy terms, differences among improved booster systems are below 8% in the locations analysed. In Total Equivalent Warming Impact (TEWI) terms, booster architectures get the lowest values with small differences between improved boosters.

Keywords: R744 transcritical booster; subcritical booster; cascade; parallel compression; ejector; commercial/retail refrigeration

---

1. Introduction

Commercial refrigeration systems are large contributors to the Greenhouse Effect due to four aspects: large refrigerant charges, high leakage rates, high energy consumption, and use of high GWP refrigerants such as R404A (GWP = 3922) or R507A (GWP = 3985) [1,2]. Europe, leading the fight against the climate change, has adopted a restriction for centralised refrigeration systems in the commercial sector from 1 January 2022 on (Regulation (EU) No 517/2014 [3]). This regulation has limited the GWP of the refrigerant to 150, in multipack centralised refrigeration systems for commercial use with capacity of 40 kW or more, with the exception of the refrigerant for the primary circuit of cascade systems where the GWP limit of 1500 has been considered. To accomplish this restriction, direct expansion in centralised refrigeration systems must rely on refrigerants with GWP less than 150. Among the different possible fluids to be used, the most suitable fluids are R744, HFO synthetic refrigerants and their mixtures, and HC, as it has been studied by members of the British Refrigeration Association et al. [4].

Supermarket refrigeration in Europe was dominated by direct expansion systems with R404A for low and medium temperature [5] and recently by R134a/R744 cascade which was experimentally analysed by Sanz-Kock et al. [6]. However, due to the restrictions established by the F-Gas, some of the existing commercial refrigerating plants will disappear as they cannot be reconverted to completely fulfill the F-Gas Regulation. In many cases, installations will have to modify their technology in order to avoid refrigerants with GWP greater than 150 to direct expansion. One possible solution for the existing systems is to replace the direct expansion systems by secondary fluid loops, such as the ones described by Wang et al. [7]. Nonetheless, the introduction of secondary fluid loops generally introduces penalties in the energy efficiency of the system. Sánchez et al. [8], measured energy increments up to 14.0% when reconverting a R134a/CO_2 direct cascade to an indirect one using propylene-glycol at medium temperature. Llopis et al. [9], presented an experimental evaluation of energy consumption reconverting a direct expansion refrigeration system to an indirect system, using R134a and R507A, with energy consumption increases up to 22.8% (R134a) and 38.7% (R507A).

Another possibility for accomplishing the F-Gas Regulation is to use systems whose operation rely on R744 as the main refrigerant. Shecco presented a report [10] of the effect that the F-Gas has on the HVAC&R industry. This report shows the number of supermarkets in Europe using transcritical systems. There were 8732 supermarkets using R744 transcritical systems in 2016, but the most part of them (98.16%) are located in northern countries where the environment temperatures are low or moderate. In cold or moderate regions, the R744 basic booster system is the commonly chosen solution since it is able to provide high energy performance [11], however, in warm regions such as Spain, Italy, or Portugal, the R744 basic booster does not offer an acceptable energy efficiency, thus in the last years, cascade systems with R744 as low temperature refrigerant have been widely used in warm regions due their good performance, but the efforts in recent years have been directed to optimising the performance of booster systems with control strategies [12], to integrate heating and air conditioning [13–15], and to improve the refrigeration architecture using advanced R744 booster systems [16] in warm regions.

Some of these improvements are: parallel compression [17] to reduce the mass flow rate of the medium temperature compressor (MTC), mechanical subcooling [18,19] to increase the specific cooling capacity in evaporators, systems with ejectors and parallel compression [20,21] to reduce exergy losses in the expansion processes and the MTC mass flow rate, and overfeed evaporators [22,23] to achieve higher evaporation temperature. In addition, the heating and air conditioning integration in the booster architecture is possible and it is an efficient solution to reduce annual energy consumption as analysed by Karampour et al. [15] and Hafner et al. [24]. Ge et al. [25] analysed a tri-generation using the high rate of heat rejection in CO_2 refrigeration systems in supermarkets to provide space cooling or refrigeration and Polzot et al. [26] analyses the energy saving potential using a Water Loop Heat Pump system. Generally, all these improvements try to reduce the power of the MTC or to use the excess thermal energy, however these systems have more inherent complexity and economic cost than an R744 basic booster.

Some researchers analysed the R744 basic booster and cascade system to compare these types of architectures in different regions with different climates. Sawalha et al. [27,28] evaluated theoretically different CO_2 centralised system solutions with two- or single-stage compression systems considering the possibility of using flooded evaporators. He compared the systems taking a R404A system and the NH_3/CO_2 cascade as reference [29] and he concluded that the cascade system is better for hot climates and the two-stage centralised systems with CO_2 are better for cold climates. Also, he indicated that the cascade system offers the best results for high temperature environments. Additionally, Fricke et al. [30] contrasted the energy consumption of cascades and booster systems in different climate regions in the USA, concluding that cascade systems obtain the lowest energy consumption in hot regions and booster systems in cold regions, however, current cascades will not be compatible with the F-Gas Regulation if the medium temperature level is served with direct expansion systems with refrigerant GWP greater than 150.

This work aims to energetically and environmentally compare several refrigeration architectures using R744 as the main refrigerant by introducing two new alternative booster arrangements working in subcritical conditions. A detailed thermodynamic analysis is presented for each configuration using a classic R513A/R744 cascade system as reference due to its extensive use in warm regions for centralised commercial refrigeration. The novelty of the analysis makes emphasis on the operational limits of the main components: compressors, expansion valves, and gas ejectors, as well as the limitations of each architecture depending on the environmental temperature. Additionally, the displacement of each compressor and the maximum operating pressure of each configuration are compared and discussed for a medium-size supermarket typically used in Spain and Portugal.

2. Refrigeration Architectures

This section presents the different refrigeration architectures considered in this work. All the architectures provide service simultaneously to medium and low temperature services using high safety refrigerants (A1). All architectures, with the exception of the R513A/R744 cascade, are in agreement with the restrictions established by the F-Gas Regulation. The R513A/R744 cascade (2.1), R513A cascaded R744 subcritical booster (2.5), and R513A cascaded R744 subcritical booster with parallel compression (2.6) always operate in subcritical conditions, either in the R513A cycle or in the R744 cycle. These cycles have been analysed using R513A (GWP = 629) as refrigerant in the high temperature cycle (HT cycle), because it is a drop-in of the widely used refrigerant in that cycle, R134a (GWP = 1430) [2], that offers improvement in COP with a reduced GWP refrigerant [31,32].

On the other hand, the other architectures analysed (basic booster (2.2), booster with parallel compression (2.3) and booster with gas ejectors and parallel compression (2.4)), use R744 as refrigerant, operating both in transcritical or subcritical conditions depending of the heat rejection temperature.

2.1. R513A/R744 Cascade Cycle (CC)

Cascade configuration, shown in Figure 1, has two independent refrigerant circuits coupled through the cascade heat exchanger (CHX). It uses R513A in the high temperature (HT) circuit and R744 in the low temperature (LT) circuit. It incorporates an internal heat exchanger (IHX) before the LT compressors (LT$_C$) to ensure the expansion of liquid in the expansion valves and a small superheating before LT$_C$, it also incorporates a desuperheater at the discharge of the LT$_C$ [33] to perform heat rejection before the cascade heat exchanger. This system was analysed theoretically by Llopis et al. [34] and investigated experimentally by Sanz-Kock et al. [6] with R134a as HT refrigerant as a solution to commercial refrigeration in warm climates. On the other hand, this architecture was also analysed with different A3 refrigerants by Sachdeva et al. [35].

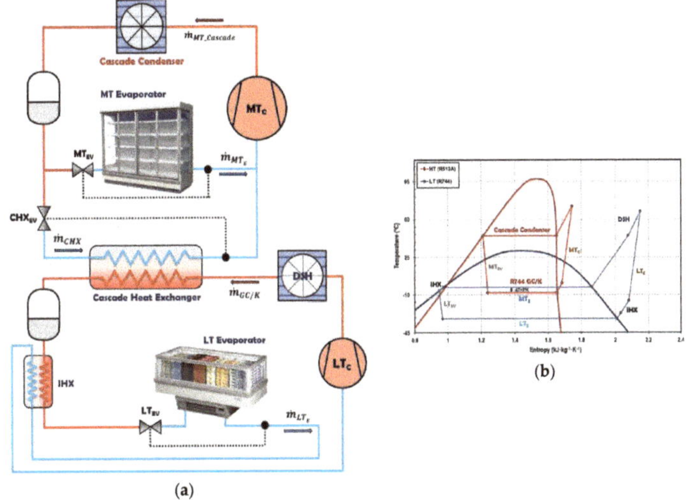

Figure 1. Schematic of a R513A/R744 cascade cycle (a) and its T–s diagram (b).

2.2. Basic Booster (BB)

The basic booster configuration, Figure 2, uses only R744 as refrigerant. It incorporates LT compressors, desuperheater (DSH), MT compressors (MT$_C$), a high pressure control valve (HPCV) to control the heat rejection pressure in order to operate in optimal conditions, a flash gas valve (FG$_V$), and a gas-cooler/condenser (GC/K) which needs to be well designed and controlled as analysed by Ge et al. [36] and Tsamos et al. [37,38]. R744 is distributed to the services using direct expansion systems. R744 is sent as saturated liquid to the MT cabinets and subcooled (due to IHX) to the LT cabinets. This system will operate in subcritical or transcritical mode depending on the environment temperature (Section 3).

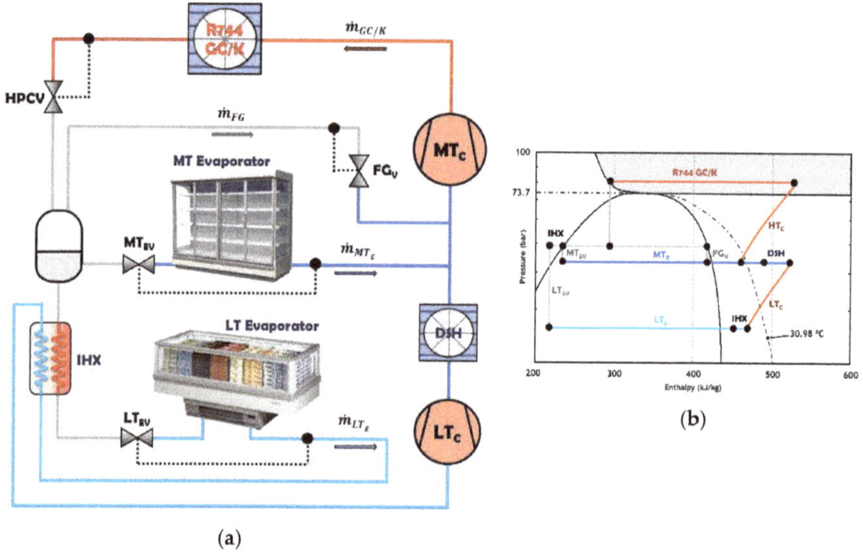

Figure 2. Schematic of a basic booster (**a**) and its p–h diagram (**b**).

2.3. Booster with Parallel Compression (BB + PC)

The booster with parallel compression configuration, Figure 3, uses the same architecture as the basic booster (BB) but it includes additional compressors denoted as parallel compressors (PC). The PC extracts saturated vapour from the flash tank and it is compressed to the high rejection pressure. As analysed theoretically by Sarkar et al. [39] in a simple stage, the use of PC allows one to reduce the pressure in the flash tank and to increase the specific cooling capacity. In addition, these compressors reduce the mass flow rate in MT and LT services and in the LT$_C$ and MT$_C$. Chesi et al. [40] also analysed this cycle, reaching COP improvements of over 30% with respect to the basic cycle and showed three main parameters with high influence on the cycle; compressors volumetric flow ratio, intermediate pressure, and the separator efficiency. BB + PC but without DSH was theoretically investigated by Gullo et al. [17], who indicated that it allows COP increments compared to the basic booster system up to 33% for an environment temperature of 35 °C. Another study by Tsamos et al. [41] showed that parallel compression is an energy efficient system for moderate and warm climates.

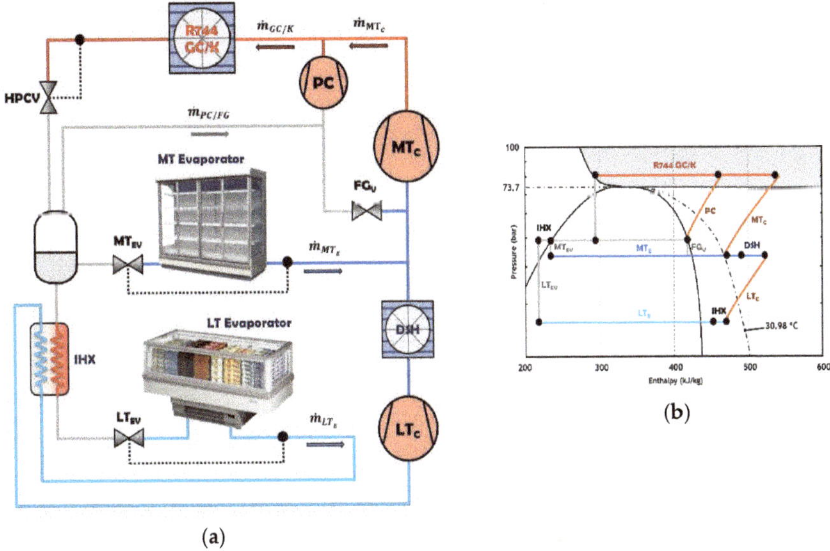

(a)

Figure 3. Schematic of a booster + PC (**a**) and its p–h diagram (**b**).

2.4. Booster with Parallel Compression and Gas Ejectors (BB + PC + GEjs)

The booster system with gas ejectors and parallel compression, Figure 4, incorporates ejectors in combination with parallel compressors. Ejectors reduce the mass flow rate through the medium temperature compressors [42] and reduce the exergy losses incrementing the specific cooling capacity in the evaporators.

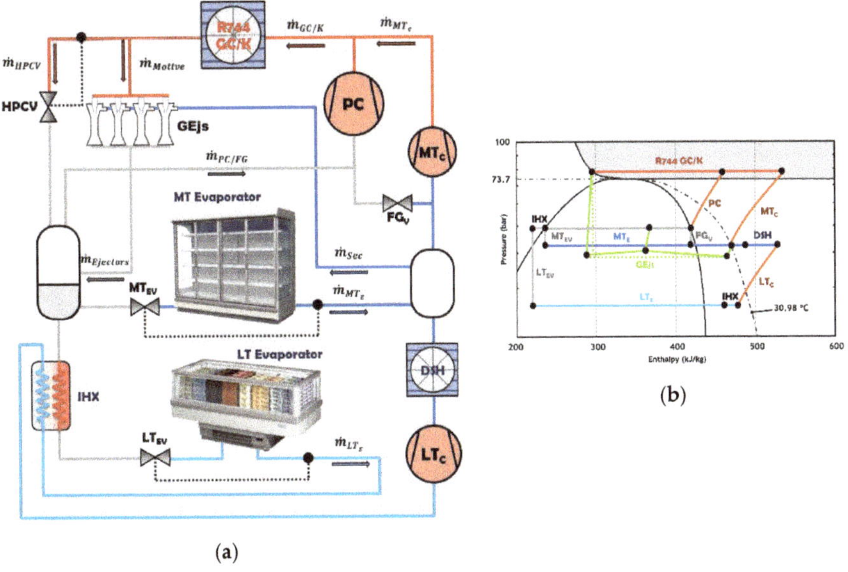

(a)

Figure 4. Schematic of a booster + PC (**a**) and its p–h diagram (**b**).

If the ejector is modulating, it could also replace the high pressure control valve of the system [21]. The use of ejectors in the R744 booster is one of the improvements with better future prospects, such as the multi-ejector system as presented by Hafner et al. [20] and by Hafner and Banasiak et al. [43].

Operation of the ejectors depends mainly on the input pressures and mass flow ratios, as analysed by Li and Groll et al. [44] and Liu and Groll et al. [45] and on the ejectors geometry as evaluated by Liu and Groll et al. [46]. Accordingly, the improvements using ejectors are very dependent on the operating conditions of the system. Nonetheless, this architecture presents COP increments respect to the booster system with parallel compression up to 7%, as analysed by Haida et al. [42].

2.5. Cascade with Subcritical Booster (CC + SB)

The next configuration is a cascade cycle, with R513A in the primary cycle (HT cycle) and a R744 subcritical booster (Figure 5). This system joins two types of architectures (cascade and booster) to obtain great performance at low and high environment temperatures, because basic boosters architecture have great performance at low environment temperatures, as does the cascade architecture at high environment temperature.

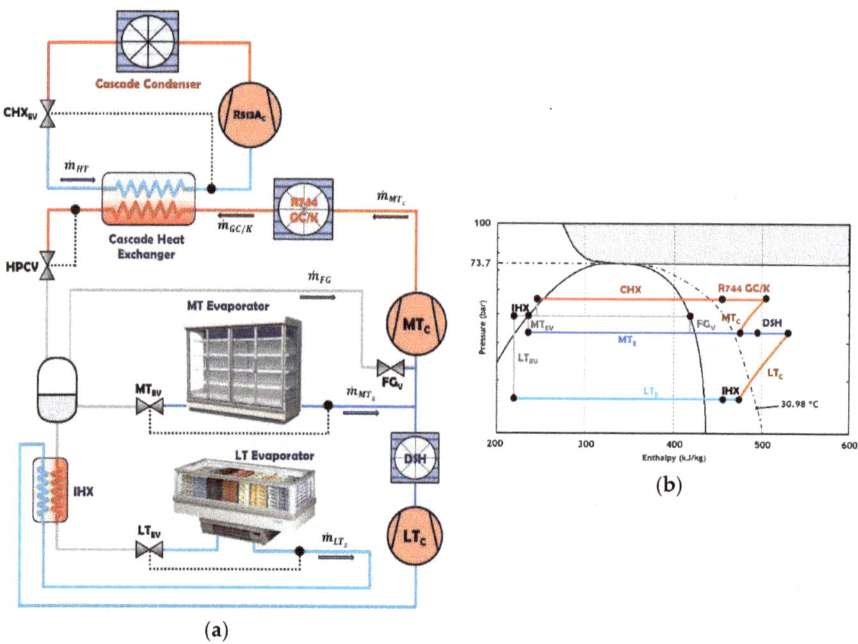

Figure 5. Schematic of a cascade with subcritical booster (**a**) and its R744 p–h diagram (**b**).

This architecture provides service simultaneously to MT and LT appliances (Section 2.2) with R744, such as the other R744 transcritical boosters, but in this case, the R744 booster always works in subcritical conditions (forced by the HT cycle).

Since MT$_C$ always work in subcritical conditions, subcritical R744 compressors can be used in this rack. The advantage is that subcritical compressors present higher efficiencies than transcritical ones operating in subcritical conditions according to the manufacturer-provided data. For this reason, the type of compressors selected for MT$_C$ in this architecture are subcritical. This is an advantage in relation to transcritical boosters working in subcritical conditions.

Another advantage of this architecture for some supermarkets with HFC/R744 cascades (widely used in warm climates), if the facility has not reached its end-life use, some parts of the existing system (condenser and HT compressors) could be used in the reconversion procedure.

This system can operate with other refrigerants in the HT cycle such as R152a, R290, and R1270, but these are not considered in this work due to their lower safety classifications.

2.6. Cascade with Subcritical Booster and Parallel Compression (CC + SB + PC)

The last configuration is similar to the previous one but uses parallel compression (PC) (Figure 6). This system corresponds to a cascade cycle with R513A in the primary cycle, maintaining the R744 booster with parallel compression in subcritical conditions. As in the previous case, this system can reuse the HT compressors and condenser of a previous cascade architecture.

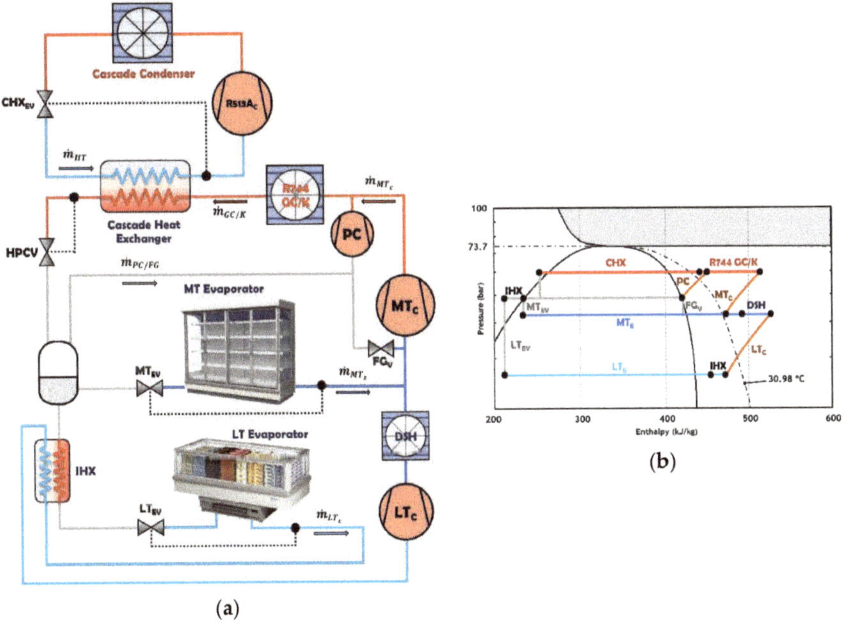

Figure 6. Schematic of a cascade with subcritical booster + PC (**a**) and its R744 p–h diagram (**b**).

For CC + SB + PC, MT$_C$ always work in subcritical and now also the PC, so these compressor racks are composed of subcritical R744 compressors.

3. Thermodynamic Modelling and Component Limitations

This section describes the thermodynamic models of each refrigeration cycle component, establishes the transition methodology between operation in transcritical and subcritical conditions, describes the reference supermarket application for sizing the systems, and shows the working limitations of each element that condition the operating of the systems.

3.1. Heat Rejection Characteristics

For R513A cycles, condensing temperature is evaluated with Equation (1), considering an environment-condensing temperature difference (ETD) of 5 K [35].

$$T_{K.R513A} = T_{env} + ETD \tag{1}$$

For R744 cycles in subcritical conditions, condensing temperature is computed with Equation (2), considering an approach temperature (ETD) and a subcooling degree in condenser (Sub). The approach temperature between the condensation and environment temperature has been fixed to 5 K too. On the other hand, a subcooling of 2 K has been considered because the proper operation of this system relies on a certain subcooling degree, as described by Danfoss [47].

$$T_{K.R744} = T_{env} + Sub + ETD \tag{2}$$

For R744 cycles in transcritical conditions, the gas-cooler outlet temperature is calculated with Equation (3), considering an approach temperature of 2 K due to the high thermal effectiveness of the gas-cooler in transcritical conditions, as measured by Sánchez et al. [48]. For these systems, the results are expressed for the optimum heat rejection pressure, which is evaluated through an iterative method.

$$T_{GC.R744} = T_{env} + ETD \tag{3}$$

3.2. Cascade Heat Exchanger (CHX)

The architectures with R513A in an HT cycle has been analysed considering a temperature difference of 5 K between the R744 condensation level and the R513A evaporation level, which it is an average value measured by Sanz-Kock et al. [6] in an experimental cascade plant.

3.3. Desuperheater (DSH)

Although not generally considered in R744 booster systems, we include a DSH in all the cycles, placed at the exit of the LT_C, since the compressor's discharge temperature is always higher than the environment temperature. This desuperheater performs heat rejection before the second compression stage, thus reducing the temperature at the MT_C inlet. This element always increases the energy efficiency of the plant. This improvement was analysed by Sanz-Kock et al. [6] in a cascade system. It has been evaluated that the introduction of a desuperheater after the first compression stage reduces energy consumption around 3–4% for each system analysed.

The outlet temperature of R744 at the DSH is evaluated with Equation (4), considering an approach of 5 K (ETD) with the environment temperature (T_{env}).

$$T_{DSH_{out}} = T_{env} + ETD \tag{4}$$

3.4. Refrigeration Heat Loads

All the refrigeration architectures have been evaluated considering the same heat load design condition, which represents a medium-sized supermarket working at standard evaporating levels [49], with cooling capacity of 41 kW and 140 kW for low and medium temperature service, respectively.

The heat load profile considered for all architectures only takes into account variations of the heat load factor between the opening and closing schedule of the supermarket, since they are always maintained at a constant inner temperature. One hundred percent heat load factor has been considered from 7:00 a.m. to 10:00 p.m. and 50% from 10:00 p.m. to 7:00 a.m. for every day of the year.

All the data described above is summarised in Table 1.

3.5. Internal Heat Exchanger (IHX)

The compressor manufacturer specifies a minimum lubricant working temperature to operate with suitable viscosity [50]. Accordingly, it is necessary to introduce an IHX in the suction line of the LT compressors. This element ensures liquid refrigerant at the inlet of the expansion valve (increasing the specific cooling capacity in evaporator) and it allows increasing the refrigerant temperature at the compressors inlet. The combination of both effects provides a small energetic improvement to the system as discussed by Llopis et al. [51]. Taking into account the experimental results from the reference, a constant thermal effectiveness of the IHX of 65% has been considered for all the refrigeration architectures.

Table 1. Summary data table.

<table>
<tr><td></td><th>CC</th><th>CC + BB</th><th>CC + BB + PC</th><th>BB</th><th>BB + PC</th><th>BB + PC + GEjs</th></tr>
<tr><td>Condensing Temp. (R513A) (°C)</td><td colspan="3">$T_{env}+5$</td><td colspan="3">-</td></tr>
<tr><td>Condensing Temp. (Subcritical R744) (°C)</td><td>−3</td><td>10</td><td>14.3</td><td colspan="3">$T_{env}+5$</td></tr>
<tr><td>GC$_{out}$ Temp. (Transcritical R744) (°C)</td><td colspan="3">-</td><td colspan="3">$T_{env}+2$</td></tr>
<tr><td>Condenser Subcooling (R513A) (°C)</td><td colspan="3">0</td><td colspan="3">-</td></tr>
<tr><td>Condenser Subcooling (Subcritical R744) (°C)</td><td>0</td><td colspan="5">2</td></tr>
<tr><td>DSH$_{out}$ Temp. (Subcritical R744) (°C)</td><td colspan="6">$T_{env}+5$</td></tr>
<tr><td>LT$_E$ Temp. (°C)</td><td colspan="6">−32</td></tr>
<tr><td>MT$_E$ Temp. (°C)</td><td>−8</td><td>5</td><td>9.3</td><td colspan="3">-</td></tr>
<tr><td>LT$_E$, MT$_E$ and CHX superheating (K)</td><td colspan="6">5</td></tr>
<tr><td>Evaporators$_{out}$ useless superheating (K)</td><td colspan="6">5</td></tr>
<tr><td>LT Load (kW)</td><td colspan="6">41</td></tr>
<tr><td>MT Load (kW)</td><td colspan="6">140</td></tr>
<tr><td>Flash Tank Limit Pressures (bar)</td><td>-</td><td>35</td><td>35–48.5</td><td>35</td><td>35–48.5</td><td>35–48.5</td></tr>
</table>

3.6. Compressors

Compressors used in this study have been modelled using data from manufacturer BITZER. Models are based on parametric adjustments of the volumetric and global efficiencies as detailed by Equations (5)–(7), whose coefficients are presented in Appendix A. Equation (6) expresses the global efficiency for all the compressors except MT$_C$ in subcritical conditions, for which the global efficiency is modeled by Equation (7), because it get a smaller error than with Equation (6). More information about these expressions in a hermetic compressor can be found in Sanchez et al. [52].

$$\eta_{volum.} = a_0 + a_1{\cdot}P_{suc} + a_2{\cdot}P_{dis} + a_3{\cdot}\left(\frac{P_{dis}}{P_{suc}}\right) + a_4{\cdot}v_{suc} \tag{5}$$

$$\eta_{global} = b_0 + b_1{\cdot}P_{suc} + b_2{\cdot}P_{dis} + b_3{\cdot}\left(\frac{P_{dis}}{P_{suc}}\right) + b_4{\cdot}v_{suc} \tag{6}$$

$$\eta_{global} = c_0 + c_1{\cdot}\left(\frac{P_{dis}}{P_{suc}}\right) + c_2{\cdot}\left(\frac{P_{dis}}{P_{suc}}\right)^2 + c_3{\cdot}\left(\frac{P_{dis}}{P_{suc}}\right)^3 \tag{7}$$

In the architectures, there are two types of compressors (subcritical and transcritical). In addition, the transcritical compressors can work in both transcritical and subcritical mode. The compressors chosen for LT$_C$ for all systems are subcritical compressors.

The MT$_C$ and PC chosen for transcritical boosters (BB, BB + PC, BB + PC + GEjs) are transcritical compressors and they can operate in transcritical or subcritical mode, thus these compressors are modelled with different equations depending on their operating mode. The MT$_C$ and PC chosen for subcritical boosters (CC + SB, CC + SB + PC) are subcritical compressors and they can operate only in subcritical mode, thus these compressors have better efficiency in subcritical mode than transcritical compressors working in subcritical mode.

3.7. Ejectors

To model the architecture operating with gas ejectors, the thermodynamic model described by Liu and Groll et al. [45] has been taken as reference. It considers the mass, movement, and energy conservation equations in each ejector part. The behaviour of each part has been modelled considering a constant efficiency of 0.8 in the motive nozzle, suction nozzle, and diffuser, as recommended by Elbel and Hrnjak et al. [53]. In addition, a constant mass entrainment ratio (ratio of suction mass flow rate respect to motive mass flow rate) of 0.25 has been considered, as an average value of mass entrainment ratios analysed by Banasiak et al. [54], for the entire operating range.

3.8. Components' Operating Restrictions

To be as close as possible to reality, the real elements' operating restrictions have been included in the analysis of the refrigeration architectures to evaluate their energy performance over a wide range of environment temperatures.

The most important operating restrictions are:

- A minimum pressure difference (ΔP) of 3.5 bar is needed in all the expansion valves for proper operation and regulation [55].
- The maximum ΔP in the expansion valves is 35 bar, for proper operation and regulation [56].
- Compressors must always operate with a compression ratio (t) higher than 1.5, because at lower values the compressors can be damaged [57].
- Pressure at the inlet of the PC must always to be lower than 55 bar. This is a restriction from the compressors manufacturer.

3.9. R744 Flash Tank Pressure

The flash tank has a maximum and minimum pressure limitation. The minimum pressure limitation is needed to keep a minimum pressure differential at the expansion valves. Although a minimum pressure differential around 3.5 bar is needed in all the valves for proper operation [54], in a transcritical booster, the minimum flash tank pressure considered is approximately 5 bar higher than the pressure of the medium temperature loads [58]. Thus, for the evaporation temperature of $-6\,^{\circ}$C ($\approx$30 bar), the minimum flash tank pressure considered will be 35 bar. This limitation is considered for all booster system analysed.

The maximum pressure in the flash tank is limited by the maximum operational pressure difference (MOPD) of the LT_{EV} (35 bar) [56]. Thus, for a low temperature evaporation of $-32\,^{\circ}$C ($\approx$13.5 bar), the maximum flash tank pressure will be 48.5 bar. A similar restriction was considered by Pardiñas et al. [59].

For booster cycles without PC but with flash gas (BB and CC + SB), the receiver pressure will be kept constant at 35 bar because it has been proven that the system COP increases with lower receiver pressure. Ge and Tsou et al. [60] reached the same conclusion for a booster without PC. On the other hand, it has been proven that for booster cycles with PC (BB + PC, BB + PC + GEjs, and CC + SB + PC), the COP is not always higher with lower pressure in the receiver, mainly due to the variable efficiency of the compressor depending on the operating conditions. For this reason, the receiver pressure at architectures with PC can vary between both pressure limitations considered in systems with PC (35–48.5 bar). In addition, it has been proven that for the system with ejectors, the pressure of the receiver also depends on the efficiency of the ejectors, mass entrainment ratio, and the pressure of the gas-cooler, although the receiver pressure limitation will be the same (35–48.5 bar). Flash tank limit pressures for all systems analysed are summarised in Table 1.

3.10. Transition Zone

For transcritical booster architectures, transition between transcritical and subcritical conditions depends on the control algorithm implemented in the regulation device [61]. Figure 7 shows an example of this regulation.

Figure 7a shows the variation of the GC/K output point according to the different operating conditions, while Figure 7b shows the COP variation of the analysed configuration respect to the GC/K heat rejection pressure.

According to Figure 7, the systems operating in transcritical-subcritical regime have three operating zones: transcritical zone, subcritical zone, and transitional zone. The transitional zone marks the progressive transition between the two other zones, to avoid sudden changes in the operation conditions of the booster.

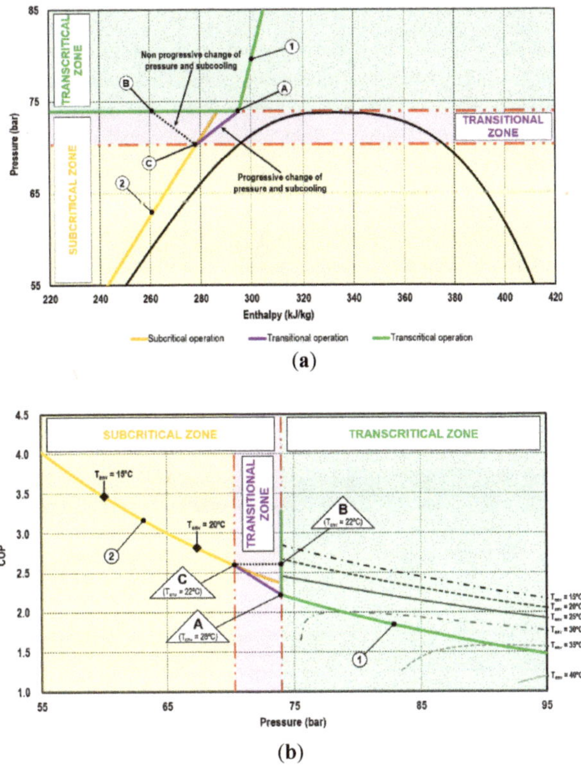

Figure 7. HPCV operation for booster ((a) (P vs. h), (b) (COP vs. P)) in subcritical, transitional, and transcritical zones.

The operation of the system in one zone or another is conditioned by the heat rejection temperature and the desired operating condition. For all the configurations analysed in this work, the maximum COP condition has been chosen for each of the dissipation temperatures. Thus, in transcritical operation, the maximum COP depends on the optimum pressure and its operating conditions. However, in subcritical operation, the maximum COP corresponds to the minimum operating pressure, which depends on the environment temperature, the efficiency of the GC/K, working as condenser, and the subcooling considered. The transition from one zone to the other implies a compromise between getting the maximum COP and a stable operation of the booster, since although the maximum COP is not achieved, the installation has a stable operation in the transition from transcritical to subcritical.

In Figure 7, for environment temperatures above 28 °C, the system follows the transcritical operation curve (line 1) with optimum GC/K pressures above the critical point. For an environment temperature about 28 °C, the optimum COP is obtained at R744 critical pressure (Point A). At this point, for lower environment temperatures, the maximum COP is achieved by keeping the GC/K pressure in transcritical operation until it reaches an environment temperature in which the system COP equals the transcritical one (Point B) as in subcritical operation (Point C). For lower environment temperatures, the highest COP is obtained with the system in subcritical operation (line 2).

In order to avoid a sudden change from transcritical (point A) to subcritical operation (point C), GC/K pressure and subcooling must be varied progressively (line A–C). Thus, the booster system operates stably but with little COP reduction.

3.11. Architectures' Working Cycles

The introduction of these restrictions to the refrigeration architectures, limit the operation of some components of the refrigeration systems over all the range of environment temperatures that they need to cover. The consequence is that a defined architecture will operate with a different cycle all over the environment temperature range, as summarised by Figure 8 and detailed for each cycle architecture in the following subsections.

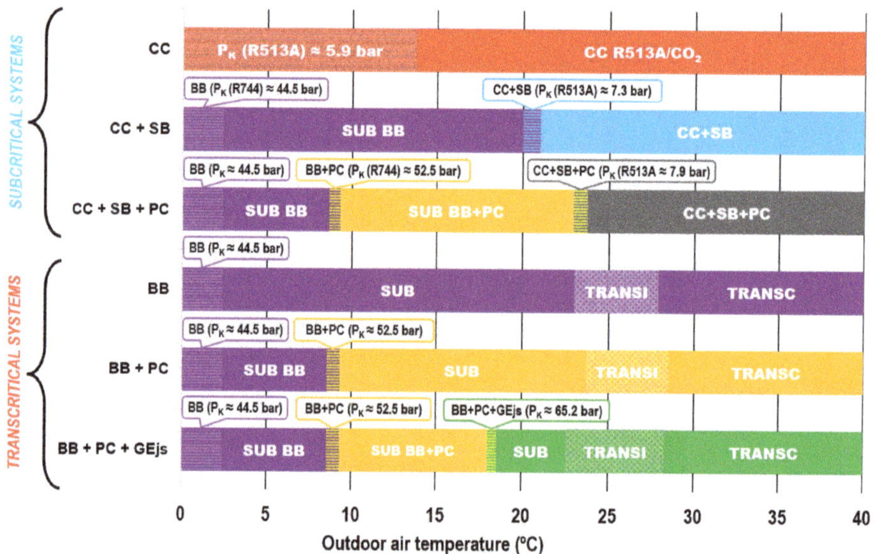

Figure 8. Operation of each architecture at different environmental temperatures.

3.11.1. R513A/R744 Cascade System (CC)

With the considered limitations for the expansion valves ($\Delta P \geq 3.5$) and compressors ($t \geq 1.5$) (Section 3.8), the cascade system operates in the R513A cycle with floating condensing pressure for outdoor air temperatures higher than 13.7 °C. For lower temperatures, the condensing pressure is maintained at 5.9 bar (18.7 °C) in order to guarantee the proper operation of the R513A expansion valve ($\Delta P = 3.5$).

3.11.2. R744 Basic Booster (BB)

As can be seen in Figure 8, the R744 basic booster cycle operates in four zones: optimised transcritical cycle for outdoor air temperatures higher than 27.9 °C (TRANSC); transition zone (TRANSI) from 27.9 °C to 23 °C as described in Section 3.10; subcritical zone (SUB) with floating condensing pressure from 23 °C to 2.4 °C; and finally, subcritical zone maintaining the condensing pressure of the heat rejection at 44.5 bar (9.4 °C), to avoid compression ratios lower than 1.5. The compression ratio restriction for the MT_C is more restrictive than the minimum ΔP at the medium temperature expansion valve (MT_{EV}).

3.11.3. R744 Booster + PC (BB + PC)

The R744 booster system with parallel compression (PC) operates similarly to the basic booster (transcritical, transition, and subcritical), but operating with the PC for outdoor air temperatures higher than 9.3 °C. Below this temperature, the compression ratio of the PC reaches 1.5, so the heat rejection pressure is kept constant at 52.5 bar ($t_{PC} = 1.5$). For temperatures below 8.6 °C, the COP of the BB + PC is lower than the BB, therefore, below this outdoor air temperature, the PC is disconnected and the system starts operating as BB, thus, from 8.6 °C to 2.4 °C, the refrigerating plant operates as a BB with floating condensing pressure, and below 2.4 °C, the heat rejection pressure is kept constant at 44.5 bar to avoid a compression ratios in MT_C lower than 1.5.

3.11.4. R744 Booster + PC + Gas Ejectors (BB + PC + GEjs)

The R744 booster system with gas ejectors and parallel compression operate in transcritical, transition, and subcritical conditions for outdoor air temperatures higher than 18.5 °C. At this temperature (18.5 °C), the flash tank pressure reaches 35 bar, so, from 18.5 °C to 18 °C, the GC/K pressure is kept constant at 65.2 bar. Below 18 °C, the COP of the BB + PC + GEjs is lower than the BB + PC, therefore, below this environmental temperature, the GEjs are disconnected and the system starts operating as BB + PC, thus for lower outdoor air temperature, the operation is the same as described for the BB + PC.

3.11.5. R513A Cascade + R744 Subcritical Booster (BB + SB)

The R513A cascade with R744 subcritical booster operates with floating condensing pressure at the HT cycle for outdoor air temperatures above 21 °C. At this temperature (21 °C), the ΔP in the HT expansion valve is of 3.5 bar, so, below this value, the HT condensing pressure is kept constant at 7.3 bar. From 21 °C to 20 °C, the pressure of the HT condenser is kept constant but for lower temperatures than 20 °C, the COP of the BB is greater than the BB + SB, therefore, below 20 °C, the facility operates as a R744 basic booster as previously described.

3.11.6. R513A Cascade + R744 Subcritical Booster + PC (BB + SB + PC)

The R513A cascade with R744 subcritical booster and parallel compression operates with floating condensing pressure at the HT cycle for outdoor air temperatures above 23.8 °C. At this temperature (23.8 °C), the compression ratio of the parallel compressors is lower than 1.5, and consequently the dissipation pressure of the R513A cycle must be kept constant at 7.9 bar. The limit in this case occurs at higher environment temperatures (23.8 °C) than the previous architecture because the use of the parallel compression raises the R744 condensing level to avoid a compression ratio of PC lower than 1.5. So, from 23.8 °C to 22.9 °C, this system operates like a cascade with fixed condensing pressure at the R513A cycle. For outdoor environmental temperatures below 22.9 °C, the COP of the BB + SB + PC architecture is lower than that of BB + PC, the R513A cycle is disconnected, and the system starts working as a booster with parallel compression (BB + PC) as described previously.

3.12. Other Considerations

In addition to the specific limitations considered in each architecture, the following considerations and assumptions are taken to obtain the energy results of all the configurations:

- Heat rejection pressures: the evaluation of the energy efficiency of each architecture is computed at the optimum heat rejection pressure corresponding to the pressure that maximises COP.
- Pressure drops and heat losses with environment are neglected.
- The ejectors are modelled as if they operated at maximum efficiency throughout the temperature range analysed.
- The motive and secondary flows have the same pressure at the ejectors mixing zone inlet.
- Kinetic energy at the inlet and outlet of the ejectors are considered as negligible.
- All the thermo-dynamic properties were evaluated using Refprop 9.1 [62] and all architectures are modelled with Matlab (2016b, Mathworks, USA) and VBA (2016, Microsoft, USA).

4. Results and Discussion

This section is devoted to presenting the results of the performance evaluation of all the architectures presented in Section 2. The systems are evaluated using the models and assumptions detailed in Section 3 for a range of environment temperatures from 0 to 40 °C.

The first part is dedicated to analysing the COP and capacity over the whole environment temperature range. For this range, the emphasis is on the evaluation of the compressor displacements and the optimum operating pressures. The second part is devoted to presenting the energy consumption of the systems with the conditions summarised in Section 3.4 in several representative cities of Spain and Portugal.

4.1. COP

The COP evolution for each refrigeration architecture, evaluated with Equation (8), over the range of environment temperatures is presented in Figure 9. The COP values are the optimum at each outdoor air temperature, considering optimised heat rejection pressure and working cycle according to Figure 8.

$$COP = \frac{Q_{O_{LT}} + Q_{O_{MT}}}{Pc_{LTc} + Pc_{MTc} + Pc_{PC} + Pc_{R513Ac}} \tag{8}$$

Compressor power is calculated using Equation (9).

$$Pc_i = \frac{\dot{m}c_i \left(h_{out_i} - h_{in_i} \right)}{\eta_{global_i}} \tag{9}$$

As it can be observed in Figure 9, for environment temperatures above 14 °C, R513A/R744 cascade (CC) presents the highest COP among all the configurations with a significant COP increase compared to transcritical boosters. However, for environment temperatures below 12 °C, all booster architectures present higher COP in relation to CC. This is mainly because at low environment temperatures, the high temperature cycle of the cascade is forced to condensate at a fixed pressure to avoid small pressure differentials in the HT expansion valve, as analysed in Section 3.8.

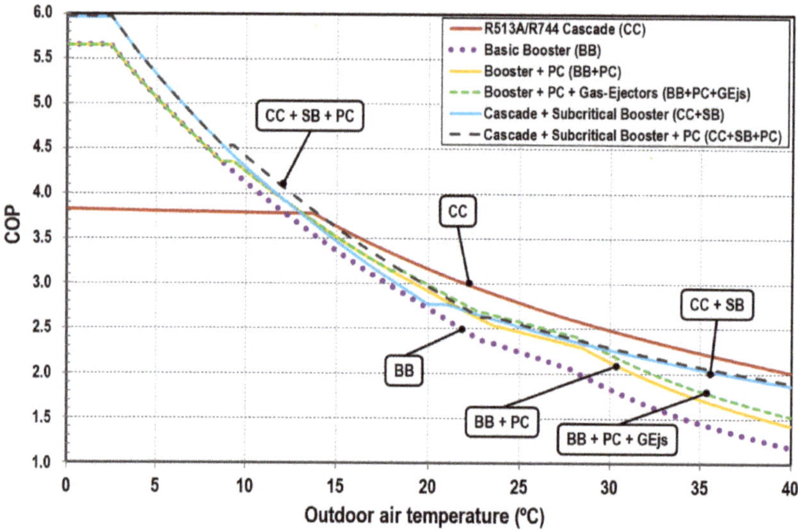

Figure 9. Systems' COP at different environment temperatures.

On the other hand, basic booster has the lowest COP of the booster systems for the whole range of analysed temperatures, but for outdoor air temperatures below 9 °C, all transcritical boosters (BB, BB + PC, and BB + PC + GEjs) have identical COP because for temperatures lower than 9 °C, these systems operate as BB, however, the subcritical boosters (CC + SB and CC + SB + PC) have higher COP than any other system analysed for temperatures below 12 °C for CC + SB and below 14 °C for CC + SB + PC. This is mainly due to the fact that these two systems use subcritical compressors which have better performance than transcritical ones working in subcritical condition. For all booster architectures, for environmental temperature of 2.4 °C, the limitation of the minimum compression ratio at MT_C is achieved, so below this temperature, the rejection pressure in the R744 cycles is kept constant.

For outdoor air temperatures between 14 and 30 °C, all improved boosters have similar COP but with an important COP improvement compared to the BB, mainly for temperatures above 20 °C. For high environment temperatures (above 30 °C), the COP differences increase between R744 boosters, with the CC + SB + PC architecture having the highest COP, followed by CC + SB with a similar COP, then the BB + PC + GEjs, BB + PC, and finally the BB.

The COP differences between each architecture compared to the R513A/R744 cascade (CC) and R744 basic booster (BB) are presented in Figures 10 and 11, respectively, and summarised in Table 2. The COP variations are obtained with Equations (10) and (11).

$$\Delta COP_{i/CC}(\%) = \frac{COP_i - COP_{CC}}{COP_{CC}} \cdot 100 \tag{10}$$

$$\Delta COP_{i/BB}(\%) = \frac{COP_i - COP_{BB}}{COP_{BB}} \cdot 100 \tag{11}$$

In relation to the R513A/R744 cascade, all booster architectures present COP improvements for environment temperatures from 0 °C to 12 °C. On the one hand, these improvements range from 48% to 0% for transcritical boosters (BB, BB + PC, and BB + PB + GEjs) with hardly any difference between them. Accordingly, these booster improvements are not justified for temperatures below 12 °C. On the other hand, cascaded subcritical boosters (CC + SB and CC + SB + PC) have a COP improvement in relation to CC from 56% to 0%, with a small COP improvement over the BB up to 5.4–7.3% for environment temperatures from 0–12 °C.

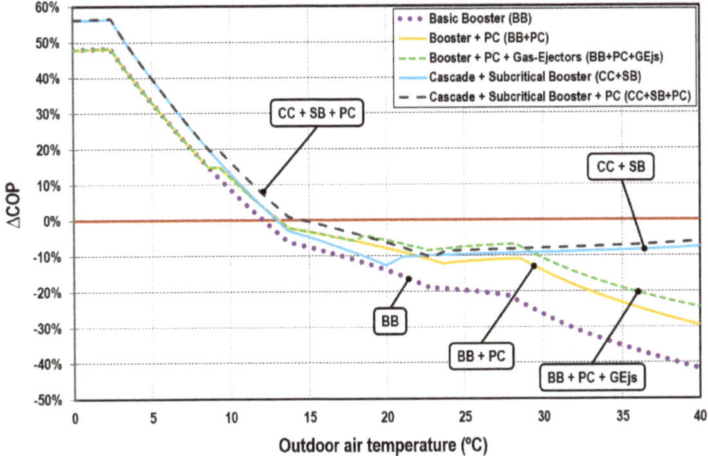

Figure 10. COP increment of each system with respect to R513A/R744 cascade.

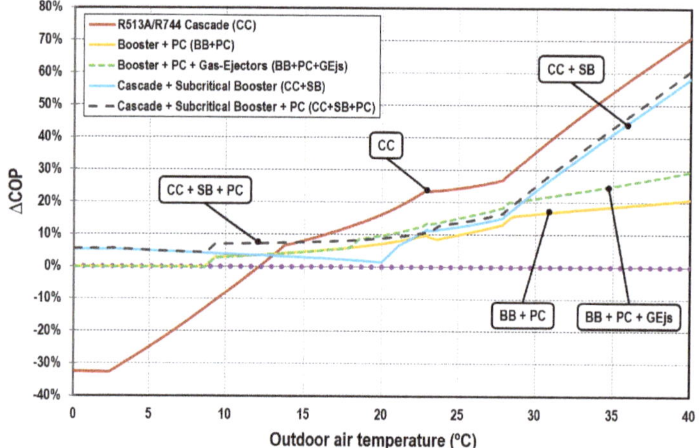

Figure 11. COP increment of each system with respect to R744 basic booster.

Table 2. COP variation of each architecture compared with the R513A/R744 cascade (CC) and R744 basic booster (BB).

T_{air} (°C)	BB		BB + PC		BB + PC + GEjs		CC + SB		CC + SB + PC	
	ΔCOP_{CC} (%)	ΔCOP_{BB} (%)	ΔCOP_{CC} (%)	ΔCOP_{BB} (%)	ΔCOP_{CC} (%)	ΔCOP_{BB} (%)	ΔCOP_{CC} (%)	ΔCOP_{BB} (%)	ΔCOP_{CC} (%)	ΔCOP_{BB} (%)
0	47.95	-	47.95	0.00	47.95	0.00	56.03	5.46	56.03	5.46
5	33.16	-	33.16	0.00	33.16	0.00	39.88	5.05	39.88	5.05
10	8.99	-	12.29	3.03	12.29	3.03	13.44	4.08	16.62	7.00
15	−7.42	-	−3.12	4.64	−3.12	4.64	−4.57	3.08	−0.43	7.55
20	−13.89	-	−7.75	7.13	−5.54	9.69	−12.51	1.60	−6.30	8.82
25	−19.53	-	−11.54	9.93	−7.45	15.02	−9.50	12.46	−8.55	13.65
30	−26.19	-	−14.07	16.42	−10.61	21.11	−8.95	23.36	−7.90	24.78
35	−35.12	-	−23.04	18.63	−18.92	24.97	−8.36	41.24	−7.18	43.07
40	−41.48	-	−29.32	20.79	−24.27	29.40	−7.41	58.23	−5.98	60.66

However, for temperatures above 14 °C, the COP of all booster architectures falls below that offered by the R513A/R744 cascade. The reductions in COP in relation to the R513A/R744 cascade solution vary from 0 to 41.5% for the BB, from 0 to 29.3% for the BB + PC, and from 0 to 24.3% for BB + PC + GEjs, whereas the reductions of cascaded R744 subcritical boosters are small; for the CC + SB architecture, the reductions vary from 0 to 7.4% and for the CC + SB + PC from 0 to 6%. These COP variations of each architecture with respect to the BB and CC are summarised in Table 1.

In booster architectures, improved transcritical boosters get COP increments over the BB, for outdoor air temperatures above 8.6 °C, up to 20.8% for BB + PC and up to 29.4% for BB + PC + GEjs, however, cascaded subcritical boosters have COP increments from 1.6 to 58.2% for CC + SB and from 5% to 60.7% for CC + SB + PC with respect to BB for whole range of temperatures analysed.

4.2. Displacement of Compressors

In this subsection, the displacement of each type of compressor rack and its variation range is presented and discussed for all the environment temperatures. This parameter allows one to select the number and the size of the compressors to be installed at the refrigerating facility. Calculations are made considering refrigeration heat loads for MT and LT established in Section 3.4. The results of the range of displacement calculated for each compressor rack are analysed graphically in the next figures (summarised in Table 3).

Table 3. Compressor displacements and their variation over all the range of environment temperatures analysed.

CYCLE	LTc (R744)				MTc (E744)				PC (R744)				HFCc (R513A)			
	MODEL	d_{min} (m³/h)	d_{max} (m³/h)	d_{var} (%)	MODEL	d_{min} (m³/h)	d_{max} (m³/h)	d_{var} (%)	MODEL	d_{min} (m³/h)	d_{max} (m³/h)	d_{var} (%)	MODEL	d_{min} (m³/h)	d_{max} (m³/h)	d_{var} (%)
BB	A1	22.4	22.4	0.0	B1	42.2	94.2	38.1								
BB + PC	A2	22.4	26.0	7.6	B2	41.0	58.1	17.2	C1	5.6	22.9	60.9				
BB + PC + GEjs	A3	22.4	24.8	5.0	B3	25.3	47.1	30.2	C2	5.6	49.9	79.9				
CC	A4	22.2	22.2	0.0									D1	383.7	586.8	20.9
CC + SB	A5	22.4	22.4	0.0	B4	42.2	61.2	18.4					D2	247.7	364.9	19.1
CC + SB + PC	A6	22.4	25.8	7.0	B5	41.0	53.3	13.0	C3	5.7	9.4	24.0	D3	212.7	302.0	17.4

The displacement of the compressors are related to the volumetric efficiency, mass flow rate, and the density of the refrigerant in the suction side. The displacements for the compressors are obtained using Equation (12).

$$\eta_{vol}(\%) = \frac{\dot{m}_{LTc}}{V_{disp} \cdot \rho(T_{suction}, P_{suction})} \tag{12}$$

The next table presents the maximum and minimum compressor displacements, as well as their variation over the environment temperature range from 0 to 40 °C. Each compressor's rack for each architecture is denoted by a letter and a number to be easily identified in the graphs. The ranges of displacement shown in Table 2 are an output result of the computation model.

4.2.1. Low Temperature Compressors (LT$_C$)

The LT$_C$ use R744 as refrigerant and the displacement for each architecture, all over the environment temperature range, is presented in Figure 12. For this compressor rack, BB + PC gets the largest displacements because the flash tank optimum pressure increases for higher outdoor air temperatures, so the enthalpy increment in LT$_E$ is reduced and, for the same cooling capacity, the mass flow rate through LT$_E$ and LT$_C$ increases. The same happens with the other systems with parallel compression (BB + PC + GEjs and CC + SB + PC).

For this type of compressor, the differences between all architectures are smaller, with a maximum displacement variation from 0 to 40 °C of ±7.6% for BB + PC, for a fixed cooling capacity of 41 kW.

Other architectures have similar LT$_C$ displacement in the whole range of temperatures because the conditions before and after the LT$_E$ are similar. In R513A/R744 cascade, the HT cycle forces the R744 cycle to work in similar conditions, and for R744 boosters without PC, the flash tank keeps a constant pressure of 35 bar.

For outdoor air temperatures above 23 °C, the CC + SB + PC operates with the minimum pressure in the flash tank (35 bar) like architectures without PC, because this system is limited by the compression ratio of the PC when operates with R513A cycle. For lower temperatures, the R513A compressor of the HT cycle is disconnected and this architecture works as SB + PC with a slight difference in displacement compared to the compressors A2 used in BB + PC architecture, because these architectures operate with a different model of compressors.

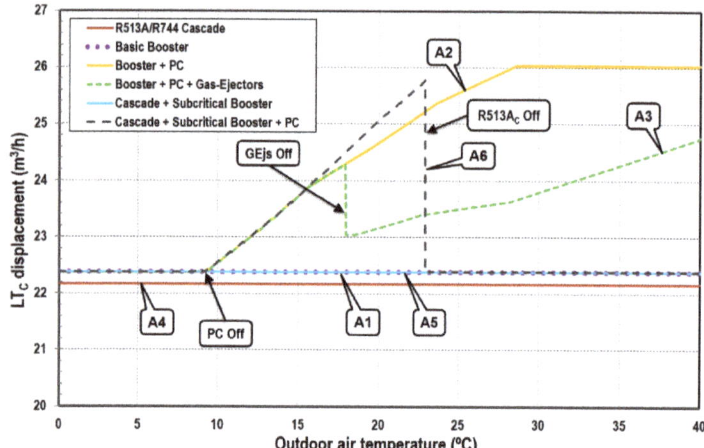

Figure 12. Low temperature compressor displacement.

4.2.2. Medium Temperature R744 Compressors (MT$_C$)

MT$_C$ displacement for each booster architecture is presented in Figure 13 for the entire environmental temperature range. The MT$_C$ with larger variation of displacement corresponds to the BB system (B1), with a maximum displacement of 94 m^3·h^{-1} and a minimum of 42.2 m^3·h^{-1}. This represents a displacement variation of 38.1%. However, improved boosters allow reductions in the MT$_C$ displacement between 38% and 73% compared to the BB, where a booster with ejectors is the system with greatest displacement reduction. For this rack of compressors, due to their displacement variations, it is necessary to use variable frequency drives in order to satisfy the cooling demand. This affects the final cost of the architectures.

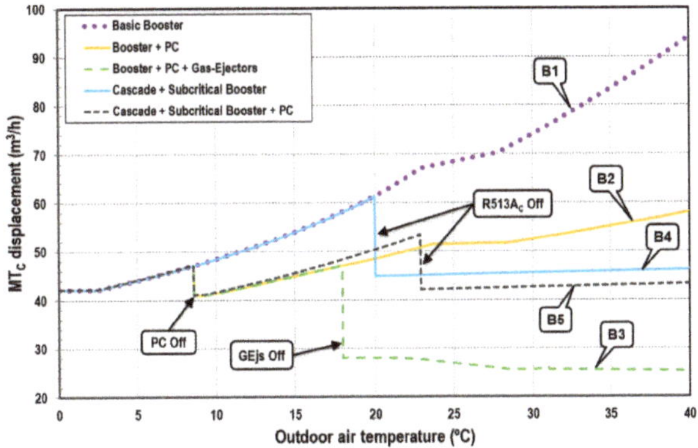

Figure 13. Medium temperature compressor displacement.

4.2.3. R744 Parallel Compressors (PC)

Parallel compressor displacements are presented in Figure 14. As can be observed for the three configurations, the range of displacement needed to cover the entire operating range (0–40 °C) is 60.9% for BB + PC system, 79.9% for BB + PC + GEjs, and 24% for PC. These large variations in displacement are important because, according to the compressor's manufacturers, the maximum variation range is generally ±40% [49]. For transcritical boosters with PC (BB + PC and BB + PC + GEjs), the variation range will force one to design the parallel compression rack using more than one parallel compressor if the facility is designed for outdoor air temperatures between 8.6–40 °C. Otherwise, for a subcritical booster (BB + SB + PC), this variation can be absorbed by only one compressor.

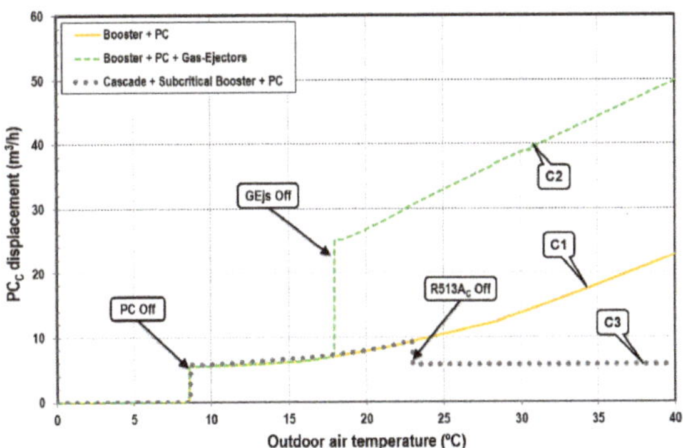

Figure 14. Parallel compressor displacement.

4.2.4. R513A Compressors

R513A compressor displacements for CC, CC + SB, and CC + SB + PC are detailed in Figure 15. It needs to be mentioned that the displacement of these compressors is of another magnitude than of the R744 ones due to the different specific volumes of R513A. The first observation that can be made is that the CC systems need practically double displacement compared to the configurations of cascaded R744 subcritical booster systems. That is because the R744 condensing level with the cascade is lower than in the other systems. In addition, the variation range of displacement for this configuration remains below 20.9% in all environments. It is important to clarify that the refrigerant R513A is more expensive than R744, so this affects to the final cost of the cascaded systems, although R744 compressors are more expensive than these.

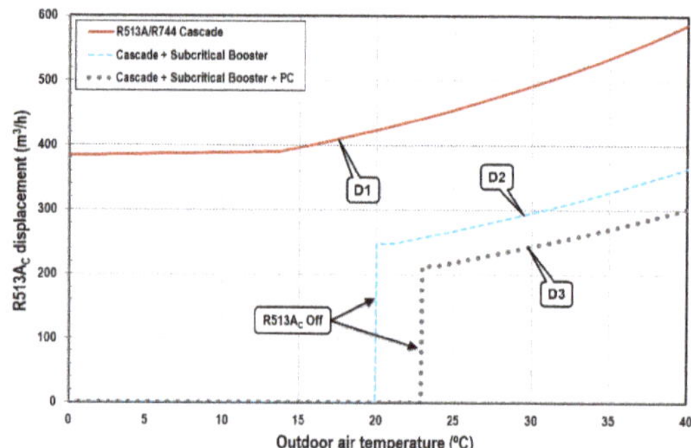

Figure 15. R513A compressor displacement.

4.3. Maximum Operating Pressures

Another important aspect to be mentioned regarding the refrigeration architectures corresponds to the maximum operating pressures, which condition the classification of the plant regarding safety. Figure 16 presents the optimum working pressures of all the compressors in all the environment temperature range, and Table 4 collects their maximum operating pressures. As it can be observed, the range of pressure needs to be considered. First, the transcritical boosters (BB, BB + PC, BB + PC + GEjs) present maximum operating pressures in the evaluated range near 100–106 bar. Next, the cascaded R744 subcritical booster systems present a maximum working pressure below 67–72 bar. Finally, the cascade solution has a maximum R744 working pressure below 35 bar.

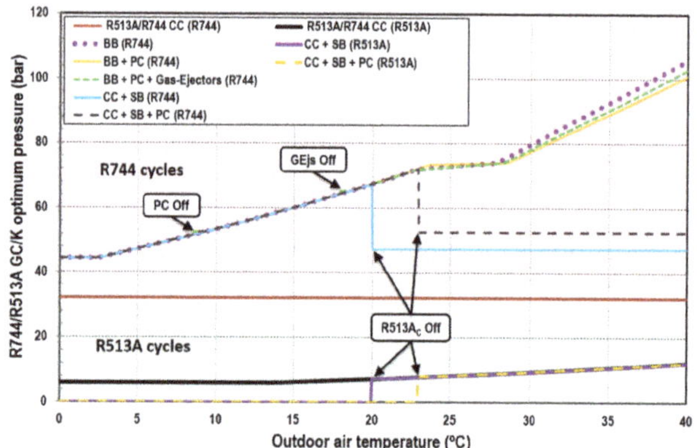

Figure 16. Optimal pressure at different environment temperatures.

Table 4. Maximum optimum pressures.

CYCLE	Max P_{opt} (R744) (bar)	Max P_{opt} (R513A) (bar)
BB	105.8	
BB + PC	100.8	
BB + PC + GEjs	102.7	
CC	32.2	12.2
CC + SB	67.4	12.2
CC + SB + PC	72.1	12.2

4.4. Annual Energy Consumption (AEC)

This section presents the annual energy consumption of the different analysed refrigeration architectures for some representative cities of Spain and Portugal. The outdoor air temperatures for each city are obtained using the EnergyPlus tool [63] to simulate the systems' hourly performance.

The energy consumption over 24 h throughout an entire year represents the estimated energy consumption of the refrigeration system for supplying the refrigeration loads detailed in Section 3.4. The heat load profile of the systems always considers the inside of the supermarket to be at the same conditions, since they are air-conditioned, and only takes into account a different heat load factor (LF) depending on the opening or closing schedule of the supermarkets. In this case, 100% heat load factor has been considered from 7:00 a.m. to 10:00 p.m. and 50% from 10:00 p.m. to 7:00 a.m. for all the days of the year. The COP of each refrigeration architecture has been considered at its optimum conditions, those detailed in Figure 9. The annual energy consumption for each location is calculated using Equation (13).

$$AEC(kWh) = \sum_{m=1}^{12} \sum_{H=1}^{24} \frac{Q_{OLT} + Q_{OMT}}{COP_{m,H}} \cdot \frac{LF}{100} \cdot D_m \tag{13}$$

Figure 17 represents the annual energy consumption of each architecture for each location, as well as their average annual temperature, and Figure 18 represents the energy consumption variation of each one, calculated by Equation (14), in relation to the R513A/R744 cascade configuration, which is considered as a reference system in warm climates.

$$\Delta E_i(\%) = \frac{E_i - E_{CC}}{E_{CC}} \cdot 100 \tag{14}$$

From the results presented in Figure 17, the following inferences can be made:

- For locations with lower annual average temperature (Burgos and León), all the analysed booster architectures offer energy consumption reductions compared to the R513A/R744 cascade. They range from 3.9% for the BB system, to 14.5% for the BB + SB + PC system. Between transcritical booster architectures, the best performing system for these locations is the BB + PC + GEjs, however the additional investment cost due to addition of parallel compressors and gas ejectors must be amortised by the energy consumption reduction, which reaches 5.3% compared to the BB system and only 1% over the BB + PC. The same happens with subcritical booster architectures at these locations, where CC + SB + PC only reaches an energy reduction of 2.1% with respect to the same system without PC.
- For locations with annual average temperature from 13 to 15 °C (Madrid, Oporto and Barcelona), all improved boosters offer similar energy consumption to the R513A/R744 system. The differences in energy consumption are below 3% for all improved boosters, except for CC + SB + PC in Madrid with an energy reduction up to 4.2%. Accordingly, the construction cost of the architecture will be the determinant conditions.
- For the locations with the highest temperatures (Lisbon, Valencia, and Seville) the best performing system is the R513A/R744 cascade. All booster architectures present higher energy consumption.

For Lisbon and Valencia, the BB presents an energy increment of 10%, the BB + PC around 4%, the BB + PC + GEjs about 2.3%, 4.2% for CC + SB, and the CC + SB + PC around 1.3%. Accordingly, for these annual temperatures, all systems except the BB would be recommended.

- Finally, in Seville, the gap in energy consumption, in relation to R513A/R744 cascade, increases, with BB + PC + GEjs and cascaded R744 subcritical boosters being the architectures with an increase in energy consumption below 5%. For this location, BB presents an energy increase of 14.2%, 6.7% for BB + PC, 4.5% for BB + PC + GEjs, 5% for CC + SB, and 2.4% for CC + SB + PC.

To summarise the energy consumption results, it can be affirmed that for locations with annual average temperature below about 13 °C, the R744 booster architectures are recommended. For intermediate annual temperatures about 13–15 °C, all systems except BB offer a similar annual energy consumption. For higher outdoor air temperatures, the cascade is clearly the best solution, from an energy point of view, to get lower annual energy consumption. For improved booster architectures, the one with the lowest annual consumption for all the locations analysed is CC + SB + PC, but focusing on the transcritical booster architectures, the one with the lowest annual consumption for all the locations analysed is BB + PC + GEjs. It needs to be said that the comparison has been made taking into account the energy consumption of the architectures, and no environmental or cost analysis has been included.

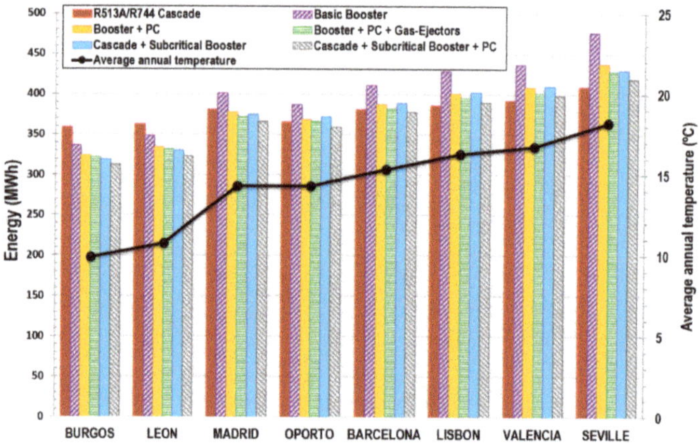

Figure 17. Annual energy consumption in different cities.

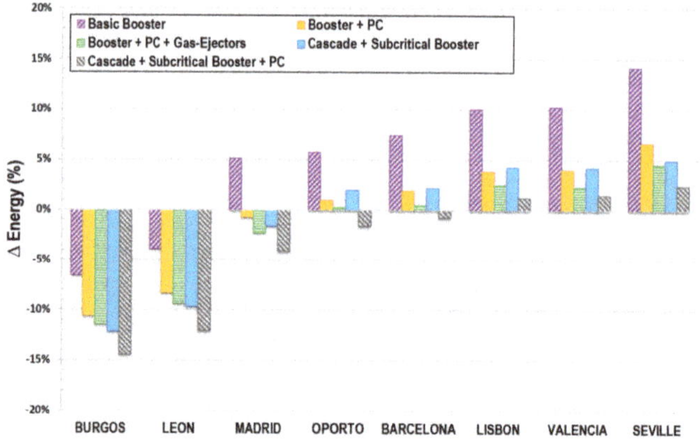

Figure 18. Energy increment with respect to R513A/R744 cascade in different cities.

4.5. Environmental Comparison

Finally, this section analyses the TEWI for all architectures evaluated in this work. This parameter includes the direct and indirect contributions to the greenhouse effect due to direct and indirect CO_2 emissions. On the one hand, the direct contribution refers to equivalent CO_2 emissions originating from the refrigerant losses (leakage and non-recovered refrigerant, Equation (15)). On the other hand, the indirect impact refers to CO_2 emissions due to energy consumption of the systems, Equation (16). Equation (17) has been used to compare the TEWI of each architecture.

$$TEWI_{Direct} = \left[GWP \cdot M_{leakage} \cdot n \right] + \left[GWP \cdot M_{ref} \cdot (1 - \alpha) \right] \tag{15}$$

$$TEWI_{Indirect} = \left[n \cdot AEC \cdot \beta \right] \tag{16}$$

$$TEWI = TEWI_{Direct} + TEWI_{Indirect} \tag{17}$$

To compare all architectures, it is necessary to consider some assumptions and so the standardised method proposed by AIRAH [64] was used as well as other assumptions adopted by Karampour et al. [65] and EMERSON Climate Technologies [49]. The basis for the TEWI evaluation is the following:

- GWPs of R513A and R744 are 629 and 1, respectively (AR4, [2]).
- Refrigeration capacity of 140 kW for MT, 41 kW for LT, and 190 kW for CC + SB and CC + SB + PC high stage evaporators.
- Refrigerant charge has been evaluated according to EMERSON Climate Technologies (2010) [49] and Karampour et al. [65]. These charges are:

  - 4 kg/kW: DX LT Centralised [49,65].
  - 2 kg/kW: DX MT Centralised [49,65].
  - 3 kg/kW: R744 Boosters (LT and MT) [65].
  - 0.75 kg/kW· high stage in CC + SB and CC + SB + PC [65].

- Annual leakage rate of 15% for all systems and 5% at high stage in CC + SB and CC + SB + PC [49,64].
- System lifetime of 10 years [49,64].
- 95% of the refrigerant charge is recovered after 10 years [49,64].
- The CO_2 emissions considered are 0.258 $kg_{CO2,equ} \cdot kWh^{-1}$ which is an annual average value for locations in the Iberian Peninsula in 2017 [66].

TEWI analysis results are shown in Figure 19. TEWI values of the reference architecture (CC) are higher than for all booster architectures for the analysed locations. This is because CC architectures use working fluids in MT with GWP values much higher than R744 and although CC has lower energy consumption than booster architectures in several locations, the direct contribution to the greenhouse gas emissions of CC is higher than for booster architectures.

Figure 20 shows the TEWI ratio of CC in relation to all booster architectures for each location analysed, which is calculated with Equation (18). A TEWI reduction from 7% to 31% is achieved with the R744 booster architectures since these systems have lower greenhouse gas emissions than the R513A/R744 cascade cycle, however, TEWI reductions between improved booster architectures are similar in the locations analysed.

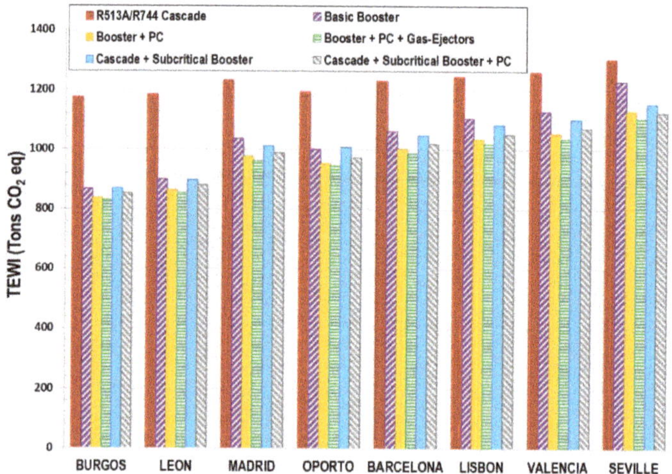

Figure 19. Total Equivalent Warming Impact (TEWI) value in the cities analysed.

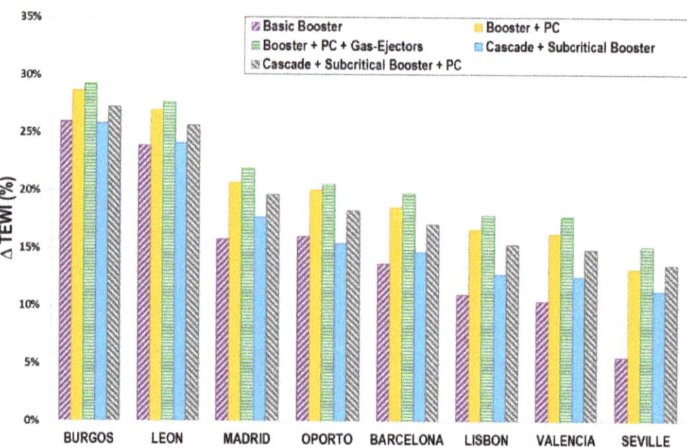

Figure 20. TEWI reductions with respect to R513A/R744 CC.

5. Conclusions

This work has analysed thermodynamically different refrigeration architectures which provide service simultaneously to medium and low temperature levels. These refrigeration architectures are in agreement with the new European Regulation 517/2014 and are candidates to replace the existing systems for supermarket refrigeration.

Systems have been evaluated using detailed thermodynamic models close to reality, where operating restrictions of ejectors, expansion valves, and compressors have been included. These restrictions determine the operating cycle of each architecture over the environment temperature range analysed (0 to 40 °C). The evaluation has covered the energetic performance of the architectures in their optimum conditions and their energy consumption in a medium-sized supermarket for different representative cities in Southern Europe.

From the performance evaluation, it has been concluded that for outdoor air temperatures below 12 °C, booster architectures present higher COP compared to CC, with COP improvements up to 48% for transcritical booster architectures (BB, BB + PC, BB + PC + GEjs) and up to 56% for subcritical booster architectures (CC + SB, CC + SB + PC). Below 9 °C, all booster architectures operate as subcritical BB. For the temperature range 9–12 °C, although improved boosters get COP increments over the BB, the COP increments between boosters are small, so for outdoor air temperatures below 12 °C, the use of these improved architectures is not entirely justified from an economic point of view.

For environment temperatures above 14 °C, R513A/R744 cascades (CC) get the highest COP among all the architectures, with COP improvements up to 41.5% over BB, 29.3% for BB + PC, 24.3 for BB + PC + GEjs, 7.4% over CC + SB, and only up to 3% with respect to CC + SB + PC.

For environment temperatures between 14 °C to 30 °C, all improved boosters have similar COP with significant COP increase regarding BB, mainly for temperatures above 20 °C. For temperatures above 30 °C, the COP between improved boosters is no longer similar, with CC + SB + PC being the booster architecture with the highest COP, followed by CC + SB with a similar COP, then BB + PC + GEjs, BB + PC, and finally BB. Subcritical boosters offer COP improvements at 40 °C compared to transcritical boosters up to 24.2% (BB + PC + GEjs), 30.2% (BB + PC), and 60.8% (BB), with the architecture with ejectors being the transcritical booster with the highest COP, with an increase up to 7.2% regarding BB + PC and up to 29.5% with respect to BB.

To counteract the cooling loads of a medium sized supermarket, taken as reference in this study, it has been observed that the R744 basic booster presents the largest R744 MT compressors, which are the ones that most affect the final consumption of the facility. In BB + PC and BB + PC + GEjs systems, the PC operates with high displacement variations, so more than one parallel compressor is needed in these boosters.

From the energy consumption analysis along a year in different locations of Southern Europe, it has been determined that for areas with annual average temperature below 13 °C, all booster architectures offer energy consumption reductions in relation to the CC. For these locations, the best performing system is CC + SB + PC and the transcritical booster with the lowest consumption is BB + PC + GEjs, with energy consumption reductions with respect to CC from 3.9% for the BB, to 14.5% for the BB + SB + PC. For locations with annual medium temperature from 13 to 15 °C, all improved boosters offer similar energy consumption to CC, with differences in energy below 3%. Finally, locations with annual medium temperature higher than 15 °C, R513A/R744 cascade is the architecture with lowest annual energy consumption, with energy reductions of 14.2% compared to BB, 6.7% for BB + PC, 4.5% for BB + PC + GEjs, 5% for CC + SB, and 2.4% for CC + SB + PC.

Finally, TEWI analysis shows that the lowest value in all locations analysed is obtained by R744 transcritical boosters with a similar value between BB + PC and BB + PC + GEjs, which is mainly due to the high charges of refrigerant with higher GWP than R744 for the architectures with R513A.

The thermodynamic analysis done in this study shows that improved boosters obtain similar energy consumptions to R513/R744 cascades for locations in warm climates, with the R513A cascaded-R744 subcritical booster with parallel compression being the booster architecture with the lowest energy consumption in the locations analysed.

Author Contributions: Methodology, J.C.-G. and D.S.; Validation, J.C.-G. and R.C.; Software, J.C.-G.; Validation, J.C.-G. and R.L.; Formal Analysis, J.C.-G. and D.S.; Investigation, J.C.-G. and L.N.-A.; J.C.-G.; Writing-Original Draft Preparation, J.C.-G.; Writing-Review & Editing, D.S., R.L., R.C.

Funding: The authors gratefully acknowledge the Spanish Ministry of Economy and Competitiveness (project ENE2014-53760-R.7) (FPI BES-2015-073612) for financing this research work.

Nomenclature

Abbreviations

AAET	annual average environment temperature (°C)
AC	air conditioning
AEC	annual energy consumption (kWh·year^{-1})
BB	basic booster
CC	cascade cycle
CHX	cascade heat exchanger
COP	coefficient of performance
d	displacement (m^3·h^{-1})
D	number of days
DSH	desuperheater
DX	direct expansion
d_var	displacement variation (%)
E	energy (kWh)
Ej	ejector
ETD	environment temperature difference (K)
GEjs	gas ejectors
FG	flash gas
GC/K	heat exchanger that works as gas-cooler or desuperheater in transcritical and as condenser in subcritical.
GWP	global warming potential
h	specific enthalpy (kJ·kg^{-1})
H	hour
HC	Hydrocarbon
HFC	hydro fluoro carbon. Referring to R513A refrigerant
HFO	hydro fluoro olefin.
HPCV	high pressure control valve
HT	high temperature
IHX	internal heat exchanger
LT	low temperature
LF	heat load factor (%)
m	month
$\dot{m}$	mass flow rate (kg·s^{-1})
M	mass (kg)
n	system lifetime (years)
MOPD	maximum operational pressure difference (bar)
MT	medium temperature
P	pressure (bar)
Pc	power consumption (kW)
PC	parallel compressors/parallel compression
$\dot{Q}$	cooling capacity (kW)
SB	subcritical booster
STD	standard temperature deviation (K)
Sub	degree of subcooling (in subcritical conditions) (K)
SUB	subcritical operation
SWEC	Spanish Weather for Energy Calculations
T	temperature (°C)
t	compression ratio
TEWI	Total Equivalent Warming Impact (ton$_{CO2,equ}$)
TRANSC	transcritical operation
TRANSI	transitional operation
v	specific volume (m^3·kg^{-1})
var	variable

$\dot{V}_{disp}$	displacement ($m^3{\cdot}s^{-1}$)
VBA	Visual Basic for Applications
Greek symbols	
α	recycling factor of the refrigerant
β	indirect emission factor ($kgCO_2{\cdot}kWh^{-1}$)
η	efficiency
Δ	increment
ρ	density ($kg{\cdot}m^{-3}$)
Subscripts	
R744	referring to R744 circuit
R513A	referring to R513A circuit
CHX	cascade heat exchanger
GC	gas-cooler or desuperheater
C	compressors
dis	discharge
env	environment
K	condenser
E	evaporator
EV	expansion valve
LT	low temperature
MT	medium temperature
O	evaporating level
leakage	refrigerant leakage
max	maximum
min	minimum
ref	refrigerant
SEC	refers to secondary flow in ejectors
suc	suction
V	valve
volum.	volumetric

Appendix A. Compressor Coefficients

The coefficients used for modelling each compressor are detailed in this appendix. They have been obtained from the real operating curves of the compressors provided by the manufacturers, both in transcritical and subcritical operation.

Using these coefficients with Equations (5)–(7), we can obtain the global and volumetric efficiency of each compressor used in this work for the entire operating range.

Compressor	Refrigerant	Compressor Type	Operating Mode	System	Coefficients			
					$\eta_{volum.}$		η_{global}	
HFC$_C$ (6FE-44Y)	R513A	Subcritical	Subcritical	CC CC + SB CC + SB + PC	a0 a1 a2 a3 a4	1.054269689 0.002487651 −0.009678312 −0.007210402 −1.10136203	b0 b1 b2 b3 b4	0.558531883 −0.0653248 0.0354848 −0.046919092 1.012723931
LT$_C$ (2ESL-4K)	R744	Subcritical	Subcritical	CC CC + SB CC + SB + PC BB BB + PC BB + PC + GEjs	a0 a1 a2 a3 a4	1.079482348 0.008315932 −0.006455476 −0.021919084 −2.55054527	b0 b1 b2 b3 b4	0.622642122 −0.00224251 0.005639549 −0.082177859 3.353632409

Point	Refrigerant			Configuration				
MT$_C$ (4FTC-20K)	R744	Transcritical	Subcritical	BB BB + PC BB + PC + GEjs	a0	0.965626672	c0	−0.302399694
					a1	0.004788087	c1	1.184852917
					a2	−0.002646382	c2	−0.46751257
					a3	−0.032284071	c3	0.060394265
					a4	−0.680966051		
			Transcritical		a0	−0.020009972	b0	0.847488488
					a1	1.092295905	b1	−0.006472439
					a2	0.001054646	b2	0.001908872
					a3	−0.002270644	b3	−0.090839418
					a4	−0.020009972	b4	5.321003429
MT$_C$ (2DME-7K)	R744	Subcritical	Subcritical	CC + SB CC + SB + PC	a0	−4.252830196	c0	−0.154129231
					a1	−0.000573377	c1	1.107677702
					a2	−0.003359172	c2	−0.460630577
					a3	−0.039423238	c3	0.06096634
					a4	-6.922851325		
PC (4MTC-10K)	R744	Transcritical	Subcritical	BB + PC BB + PC + GEjs	a0	1.064450854	b0	0.754943632
					a1	0.001960619	b1	−0.012442215
					a2	−0.001653461	b2	0.011749996
					a3	−0.073049783	b3	−0.333783866
					a4	−1.718699602	b4	17.80582359
			Transcritical		a0	1.052753205	c0	0.714231431
					a1	0.002008184	c1	−0.004470091
					a2	−0.001612749	c2	0.004286052
					a3	−0.054916991	c3	−0.183037087
					a4	−4.547670512	c4	17.57882968
PC (2GME-4K)	R744	Subcritical	Subcritical	CC + SB + PC	a0	1.333657271	b0	0.681542314
					a1	−0.001724897	b1	−0.002254769
					a2	−0.00460297	b2	0.002217279
					a3	−0.024718001	b3	−0.12237903
					a4	−9.932359991	b4	9.903497011

References

1. Francis, C.; Maidment, G.G.; Davies, G.F. An investigation of refrigerant leakage in commercial refrigeration. *Int. J. Refrig.* **2016**, *74*, 12–21. [CrossRef]
2. Solomon, S.; Qin, D.; Manning, M.; Marquis, M.; Averyt, K.; Tignor, M.; Miller, H.L.; Chen, Z. Climate Change 2007 The Physical Science Basis. Available online: http://www.ipcc.ch/pdf/assessment-report/ar4/wg1/ar4_wg1_full_report.pdf (accessed on 13 July 2018).
3. European Parliament, Regulation (EU) No 517/2014 of the European Parliament and of the Council of 16 April 2014 on Fluorinated Greenhouse Gases and Repealing. 2014. Available online: http://eur-lex.europa.eu/legal-content/EN/TXT/PDF/?uri=CELEX:32014R0517&from=EN (accessed on 7 April 2018).
4. British Refrigeration Association Action Group. Putting into Use Replacement Refrigerants. 2015. Available online: http://www.henrytech.co.uk/wp-content/uploads/2015/09/BRA-on-PURR-report_09-2015.pdf (accessed on 8 April 2018).
5. The Institute of Refrigeration; The British Refrigeration Association. The Carbon Trust, Refrigeration Road Map. 2010. Available online: https://www.epa.gov/sites/production/files/documents/refrigerationroadmap.pdf (accessed on 12 January 2018).
6. Sanz-Kock, C.; Llopis, R.; Sánchez, D.; Cabello, R.; Torrella, E. Experimental evaluation of a R134a/CO$_2$ cascade refrigeration plant. *Appl. Therm. Eng.* **2014**, *73*, 41–50. [CrossRef]
7. Wang, K.; Eisele, M.; Hwang, Y.; Radermacher, R. Review of secondary loop refrigeration systems. *Int. J. Refrig.* **2009**, *33*, 212–234. [CrossRef]
8. Sánchez, D.; Llopis, R.; Cabello, R.; Catalán-Gil, J.; Nebot-Andrés, L. Conversion of a direct to an indirect commercial (HFC134a/CO$_2$) cascade refrigeration system: Energy impact analysis. *Repos. Univ. Jaume I* **2016**. [CrossRef]
9. Llopis, R.; Sánchez, D.; Cabello, R.; Catalán-Gil, J.; Nebot-Andrés, L. Conversion of a Direct to an Indirect Refrigeration System at Medium Temperature Using R-134a and R-507A: An Energy Impact Analysis. *Appl. Sci.* **2018**, *8*, 247. [CrossRef]
10. Shecco. F-Gas Regulation Shaking up the HVAC&R Industry. 2016. Available online: http://publication.shecco.com/upload/file/org/57fe03c438c881476264900fdfko.pdf (accessed on 1 February 2018).

11. Sawalha, S. Investigation of heat recovery in CO_2 trans-critical solution for supermarket refrigeration. *Int. J. Refrig.* **2013**, *36*, 145–156. [CrossRef]
12. Karampour, M.; Sawalha, S. Performance and Control Strategies Analysis of a CO_2 Trans-Critical Booster System. In Proceedings of the 3rd IIR International Conference on Sustainability and the Cold Chain, ICCC 2014, London, UK, 23–25 June 2014; p. 8. Available online: https://www.researchgate.net/publication/281583942_Performance_and_control_strategies_analysis_of_a_CO2_trans-critical_booster_system (accessed on 18 June 2018).
13. Hafner, A.; Poppi, S.; Nekså, P.; Minetto, S.; Eikevik, T.M. Development of Commercial Refrigeration Systems with Heat Recovery for Supermarket Building. In Proceedings of the 10th IIR-Gustav Lorentzen Conference on Natural Working Fluids (GL2012), Delft, The Netherlands, 25–27 June 2012; p. 11.
14. Karampour, M.; Sawalha, S. *Supermarket Refrigeration and Heat Recovery Using CO_2 as Refrigerant*; Swedish Energy Agency: Stockholm, Sweden, 2014.
15. Karampour, M.; Sawalha, S. Energy efficiency evaluation of integrated CO_2 trans-critical system in supermarkets: A field measurements and modelling analysis. *Int. J. Refrig.* **2017**, *82*, 470–486. [CrossRef]
16. Hafner, A.; Hemmingsen, A.K. R744 Refrigeration technologies for supermarkets in warm climates. In Proceedings of the 24th International Congress of Refrigeration, Yokohama, Japan, 16–22 August 2015.
17. Gullo, P.; Elmegaard, B.; Cortella, G. Energy and environmental performance assessment of R744 booster supermarket refrigeration systems operating in warm climates. *Int. J. Refrig.* **2016**, *64*, 61–79. [CrossRef]
18. Llopis, R.; Nebot-Andrés, L.; Cabello, R.; Sánchez, D.; Catalán-Gil, J. Experimental evaluation of a CO_2 transcritical refrigeration plant with dedicated mechanical subcooling. *Int. J. Refrig.* **2016**, *69*, 361–368. [CrossRef]
19. Sánchez, D.; Catalán-Gil, J.; Llopis, R.; Nebot-Andrés, L.; Cabello, R.; Torrella, E. Improvements in a CO_2 transcritical plant working with two different subcooling systems. In Proceedings of the 12th IIR Gustav Lorenzen Natural Working Fluids Conference, Edinburgh, UK, 21–24 August 2016.
20. Hafner, A.; Försterling, S.; Banasiak, K. Multi-ejector concept for R-744 supermarket refrigeration. *Int. J. Refrig.* **2014**, *43*, 1–13. [CrossRef]
21. Gullo, P.; Hafner, A.; Cortella, G. Multi-ejector R744 booster refrigerating plant and air conditioning system integration—A theoretical evaluation of energy benefits for supermarket applications. *Int. J. Refrig.* **2016**, *75*, 164–176. [CrossRef]
22. Minetto, S.; Brignoli, R.; Zilio, C.; Marinetti, S. Experimental analysis of a new method for overfeeding multiple evaporators in refrigeration systems. *Int. J. Refrig.* **2013**, *38*, 1–9. [CrossRef]
23. Purohit, N.; Kumar Gupta, D.; Sankar Dasgupta, M. Energetic and economic analysis of trans-critical CO_2 booster system for refrigeration in warm climatic condition. *Int. J. Refrig.* **2017**, *80*, 182–196. [CrossRef]
24. Hafner, A. Integrated CO_2 system for refrigeration, air conditioning and sanitary hot water. In Proceedings of the 7th IIR Ammonia and CO_2 Refrigeration Technologies Conference, Ohrid, Macedonia, 11–13 May 2017; Available online: http://www.iifiir.org/clientBookline/service/reference.asp?INSTANCE=EXPLOITATION&OUTPUT=PORTAL&DOCID=IFD_REFDOC_0021729&DOCBASE=IFD_REFDOC&SETLANGUAGE=FR (accessed on 9 April 2018).
25. Ge, Y.T.; Tassou, S.A.; Suamir, I.N. Prediction and analysis of the seasonal performance of tri-generation and CO_2 refrigeration systems in supermarkets. *Appl. Energy* **2013**, *112*, 898–906. [CrossRef]
26. Polzot, A.; Dipasquale, C.; D'Agaro, P.; Cortella, G. Energy benefit assessment of a water loop heat pump system integrated with a CO_2 commercial refrigeration unit. *Energy Procedia* **2017**, *123*, 36–45. [CrossRef]
27. Sawalha, S. Theoretical evaluation of trans-critical CO_2 systems in supermarket refrigeration. Part I: Modeling, simulation and optimization of two system solutions. *Int. J. Refrig.* **2007**, *31*, 516–524. [CrossRef]
28. Sawalha, S. Theoretical evaluation of trans-critical CO_2 systems in supermarket refrigeration. Part II: System modifications and comparisons of different solutions. *Int. J. Refrig.* **2007**, *31*, 525–534. [CrossRef]
29. Sawalha, S.; Soleimani, A.; Rogstam, J. Experimental and Theoretical Evaluation of NH_3/CO_2 Cascade System for Supermarket Refrigeration in a Laboratory Environment. In Proceedings of the 7th IIR Gustav Lorentzen Conference on Natural Working Fluids, Trondheim, Norway, 28–31 May 2006.
30. Fricke, B.A.; Sharma, V. High Efficiency, Low Emission Refrigeration System. 2016. Available online: https://betterbuildingssolutioncenter.energy.gov/sites/default/files/attachments/Ref-system.pdf (accessed on 29 January 2018).

31. Llopis, R.; Sánchez, D.; Cabello, R.; Catalán-Gil, J.; Nebot-Andrés, L. Experimental analysis of R-450A and R-513A as replacements of R-134a and R-507A in a medium temperature commercial refrigeration system. *Int. J. Refrig.* **2017**, *84*, 52–66. [CrossRef]

32. Mota-Babiloni, A.; Makhnatch, P.; Khodabandeh, R.; Navarro-Esbrí, J. Experimental assessment of R134a and its lower GWP alternative R513A. *Int. J. Refrig.* **2017**, *74*, 682–688. [CrossRef]

33. Llopis, R.; Sanz-Kock, C.; On Cabello, R.; Anchez, D.S.; Torrella, E. Experimental evaluation of an internal heat exchanger in a CO_2 subcritical refrigeration cycle with gas-cooler. *Appl. Therm. Eng.* **2015**, *80*, 31–41. [CrossRef]

34. Llopis, R.; Sánchez, D.; Sanz-Kock, C.; Cabello, R.; Torrella, E. Energy and environmental comparison of two-stage solutions for commercial refrigeration at low temperature: Fluids and systems. *Appl. Energy* **2015**, *138*, 133–142. [CrossRef]

35. Sachdeva, G.; Jain, V.; Kachhwaha, S.S. Performance Study of Cascade Refrigeration System Using Alternative Refrigerants. *Int. J. Mech. Aerosp. Ind. Mechatron. Manuf. Eng.* **2014**, *8*, 7. Available online: http://waset.org/publications/9997619/performance-study-of-cascade-refrigeration-system-using-alternative-refrigerants (accessed on 9 April 2018).

36. Ge, Y.T.; Tassou, S.A.; Santosa, I.D.; Tsamos, K. Design optimisation of CO_2 gas cooler/condenser in a refrigeration system. *Appl. Energy* **2015**, *160*, 973–981. [CrossRef]

37. Tsamos, K.M.; Ge, Y.T.; Santosa, I.D.M.C.; Tassou, S.A. Experimental investigation of gas cooler/condenser designs and effects on a CO_2 booster system. *Appl. Energy* **2017**, *186*, 470–479. [CrossRef]

38. Tsamos, K.M.; Gullo, P.; Ge, Y.T.; Santosa, I.D.M.C.; Tassou, S.A.; Hafner, A. Performance investigation of the CO_2 gas cooler designs and its integration with the refrigeration system. *Energy Procedia* **2017**, *123*, 265–272. [CrossRef]

39. Sarkar, J.; Agrawal, N. Performance optimization of transcritical CO_2 cycle with parallel compression economization. *Int. J. Therm. Sci.* **2009**, *49*, 838–843. [CrossRef]

40. Chesi, A.; Esposito, F.; Ferrara, G.; Ferrari, L. Experimental analysis of R744 parallel compression cycle. *Appl. Energy* **2014**, *135*, 274–285. [CrossRef]

41. Tsamos, K.M.; Ge, Y.T.; Santosa, I.; Tassou, S.A.; Bianchi, G.; Mylona, Z. Energy analysis of alternative CO_2 refrigeration system configurations for retail food applications in moderate and warm climates. *Energy Convers. Manag.* **2017**, *150*, 822–829. [CrossRef]

42. Haida, M.; Banasiak, K.; Smolka, J.; Hafner, A.; Eikevik, T.M. Experimental analysis of the R744 vapour compression rack equipped with the multi-ejector expansion work recovery module. *Int. J. Refrig.* **2015**, *64*, 93–107. [CrossRef]

43. Hafner, A.; Banasiak, K. R744 ejector technology future perspectives. *7th Eur. Therm. Conf.* **2016**, *745*, 032157. [CrossRef]

44. Li, D.; Groll, E.A. Transcritical CO_2 refrigeration cycle with ejector-expansion device. *Int. J. Refrig.* **2005**, *28*, 766–773. [CrossRef]

45. Liu, F.; Groll, E.A. Analysis of a Two Phase Flow Ejector For Transcritical CO_2 Cycle. In Proceedings of the International Refrigeration and Air Conditioning Conference, West Lafayette, Indiana, 14–17 July 2008. Available online: https://docs.lib.purdue.edu/cgi/viewcontent.cgi?referer=https://www.google.es/&httpsredir=1&article=1923&context=iracc (accessed on 10 December 2017).

46. Liu, F.; Groll, E.A.; Li, D. Investigation on performance of variable geometry ejectors for CO_2 refrigeration cycles. *Energy* **2012**, *45*, 829–839. [CrossRef]

47. Danfoss. Food Retail CO_2 Refrigeration Systems. Designing Subcritical and Transcritical CO_2 Systems and Selecting Suitable Danfoss Components. 2009. Available online: http://files.danfoss.com/TechnicalInfo/Dila/01/DKRCEPAR1A102_TheFoodRetailCO2applicationhandbook_DILA.pdf (accessed on 8 April 2018).

48. Sánchez, D.; Cabello, R.; Llopis, R.; Torrella, E. Development and validation of a finite element model for water—CO_2 coaxial gas-coolers. *Appl. Energy* **2012**, *93*, 637–647. [CrossRef]

49. Emerson Climate Technologies, Refrigerant Choices for Commercial Refrigeration. 2010. Available online: http://www.emersonclimate.com/europe/Documents/Resources/TGE124_Refrigerant_Report_EN_1009.pdf (accessed on 20 January 2018).

50. Bitzer. Bitzer CO_2 Booster Rack. 2017. Available online: https://www.bitzer.de/shared_media/documentation/bao-107-1-au.pdf (accessed on 9 April 2018).

51. Llopis, R.; Sanz-Kock, C.; Cabello, R.; Sánchez, D.; Nebot-Andrés, L.; Catalán-Gil, J. Effects caused by the internal heat exchanger at the low temperature cycle in a cascade refrigeration plant. *Appl. Therm. Eng.* **2016**, *103*, 1077–1086. [CrossRef]

52. Sánchez, D.; Torrella, E.; Cabello, R.; Llopis, R. Influence of the superheat associated to a semihermetic compressor of a transcritical CO_2 refrigeration plant. *Appl. Therm. Eng.* **2009**, *30*, 302–309. [CrossRef]

53. Elbel, S.; Hrnjak, P. Experimental validation of a prototype ejector designed to reduce throttling losses encountered in transcritical R744 system operation. *Int. J. Refrig.* **2007**, *31*, 411–422. [CrossRef]

54. Banasiak, K.; Hafner, A.; Kriezi, E.E.; Madsen, K.B.; Birkelund, M.; Fredslund, K.; Olsson, R. Development and performance mapping of a multi-ejector expansion work recovery pack for R744 vapour compression units. *Int. J. Refrig.* **2015**, *57*, 265–276. [CrossRef]

55. Dick, W. *Commercial Refrigeration: For Air Conditioning Technicians*, 2nd ed.; Cengage Learning: Boston, MA, USA, 2009.

56. Danfoss. Electronically Operated Expansion Valve for CO_2 Type AKVH. 2012. Available online: http://files.danfoss.com/technicalinfo/dila/01/DKRCCPDVA1D302SmartconnectorAKVH.pdf (accessed on 23 January 2018).

57. Boudreau, P.J. The Compressor Operating Envelope. 2013. Available online: http://www.r744.com/files/the-compressor-operating-envelope.pdf (accessed on 30 November 2017).

58. Sharma, V.; Fricke, B.; Bansal, P. Comparative analysis of various CO_2 configurations in supermarket refrigeration systems. *Int. J. Refrig.* **2014**, *46*, 86–99. [CrossRef]

59. Pardiñas, Á.Á.; Hafner, A.; Banasiak, K. Novel integrated CO_2 vapour compression racks for supermarkets. Thermodynamic analysis of possible system configurations and influence of operational conditions. *Appl. Therm. Eng.* **2017**, *131*, 1008–1025. [CrossRef]

60. Ge, Y.T.; Tassou, S.A. Thermodynamic analysis of transcritical CO_2 booster refrigeration systems in supermarket. *Energy Convers. Manag.* **2011**, *52*, 1868–1875. [CrossRef]

61. Danfoss. Capacity Controller for Small CO_2 Booster Refrigeration System. 2013. Available online: http://files.danfoss.com/TechnicalInfo/Dila/01/RS8GU102_AK-PC_772_.pdf (accessed on 2 November 2017).

62. Lemmon, E.W.; Huber, M.L.; McLinden, M.O. NIST Standard Reference Database 23: Reference Fluid Thermodynamic and Transport Properties-REFPROP, Version 9.1. *Natl Std. Ref. Data Ser.* **2013**. Available online: https://www.nist.gov/publications/nist-standard reference database-23-reference-fluid-thermodynamic-and-transport (accessed on 29 January 2018).

63. EnergyPlus, DOE, BTO, NREL, EnergyPlus, (2017). Available online: https://energyplus.net/ (accessed on 12 January 2018).

64. AIRAH. Methods of Calculating Total Equivalent Warming Impact (TEWI) 2012. Available online: https://www.airah.org.au/Content_Files/BestPracticeGuides/Best_Practice_Tewi_June2012.pdf (accessed on 21 March 2018).

65. Karampour, M.; Sawalha, S. State-of-the-Art Integrated CO_2 Refrigeration System for Supermarkets: A Comparative Analysis. *Int. J. Refrig.* **2017**. Available online: https://www.researchgate.net/publication/320979787_State-of-the-Art_Integrated_CO2_Refrigeration_System_for_Supermarkets_a_Comparative_Analysis (accessed on 20 November 2017).

66. Red Eléctrica de España. CO_2 Emissions Associated to the Power Generation. 2017. Available online: http://www.ree.es/en/statistical-data-of-spanish-electrical-system/statistical-series/national-statistical-series (accessed on 18 March 2018).

Review

Multi-Ejector Concept: A Comprehensive Review on its Latest Technological Developments

Paride Gullo [1], Armin Hafner [1,*], Krzysztof Banasiak [2], Silvia Minetto [3] and Ekaterini E. Kriezi [4]

[1] Department of Energy and Process Engineering, NTNU Norwegian University of Science and Technology, Kolbjørn Hejes vei 1D, 7491 Trondheim, Norway; paride.gullo@ntnu.no

[2] Department of Thermal Energy, SINTEF Energy Research, Kolbjørn Hejes vei 1A, 7491 Trondheim, Norway; Krzysztof.Banasiak@sintef.no

[3] Construction Technology Institute, CNR National Research Council of Italy, Corso Stati Uniti 4, 35127 Padua, Italy; minetto@itc.cnr.it

[4] Danfoss A/S, Nordborgvej 81, 6430 Nordborg, Denmark; ekk@danfoss.com

* Correspondence: armin.hafner@ntnu.no

Received: 28 December 2018; Accepted: 25 January 2019; Published: 28 January 2019

Abstract: The adoption of the EU F-Gas Regulation 517/2014 and the resulting development of the multi-ejector concept have led carbon dioxide to take center stage as the sole refrigerant (R744) in several applications. Therefore, a knock-on effect on the number of supermarkets relying on "CO_2 only" refrigeration systems has been experienced. Additionally, a global consensus of commercial multi-ejector based R744 units is also intensifying as a consequence of both the promising results obtained and the other measures in force for environment preservation. Furthermore, the multi-ejector concept is expected to offer significant energy savings in other high energy-demanding buildings (e.g., hotels, gyms, spas) as well, even in warm climates. In this investigation, the evolution of R744 ejector supported parallel vapor compression system layouts for food retail applications was summed up. Furthermore, their technological aspects, the results related to the main theoretical assessments and some relevant field/laboratory measurements were summarized. Also, the experience gained in the adoption of the multi-ejector concept in transcritical R744 vapor-compression units aimed at other energy intensive applications was presented. Finally, the persistent barriers needing to be overcome as well as the required future work were brought to light.

Keywords: air conditioning; chiller; CO_2; commercial refrigeration; heat pump; heat recovery; industrial refrigeration; R744; transcritical vapor-compression system; two-phase ejector

---

1. Introduction

The implementation of the EU F-Gas Regulation 517/2014 [1] on fluorinated greenhouse gases (F-gases) has prompted the need to discontinue their use and substitute these working fluids with less environment-damaging alternatives. This holds particularly true for the sector involving high energy-demanding buildings (e.g., supermarkets, hotels, gyms), as the fulfillment of their refrigeration, cooling and heating (RC&H) needs causes significant indirect contributions to global warming as well. Furthermore, as a consequence of the ongoing sharp hydrofluorocarbon (HFC) quota reduction [1], the global refrigeration sector is facing an ever-growing shortage of "old" synthetic refrigerants (e.g., R404A, R507A) as well as a dramatic price rises in others (including new man-made working fluids, e.g., R448A, R449A, in addition to R404A, R507A and R410A) [2]. Also, in parallel with the coming into force of the F-Gas Regulation 517/2014 in Europe, a commitment on global scale to fight against climate change has been performed through the adoption and ratification of the Kigali amendment to the Montreal Protocol [3].

Carbon dioxide is perceived as a long-term working fluid for various RC&H applications [4], being non-flammable, non-toxic and offering favorable environmental, i.e., negligible Global Warming Potential (GWP) as well as zero Ozone Depletion Potential (ODP), and thermo-physical properties [5]. R744 is also inexpensive in comparison with man-made refrigerants. As suggested in [6], "CO_2 only" (or transcritical CO_2) RC&H solutions are accepted as viable and sustainable candidates in several sectors (e.g., supermarkets, vehicle air conditioning and heat pump units, domestic hot water heat pump systems and industrial applications). As examples:

- It was estimated that 9000 supermarkets relying on transcritical R744 systems were operating in Europe in 2017 [7]. However, it is expected that these units will be 25,000 in 2020 [4] and 55,000 in 2025 [4];
- The government subsidies has led "CO_2 only" heat pump units for domestic hot water (DHW) purposes to become standard in Japan;
- Transcritical R744 solutions are gaining ever-growing attraction in industrial refrigeration applications featuring large cooling loads [4].

However, the persistent challenge lies in moving "CO_2 only" vapor-compression units to warm regions worldwide. The implementation of the aforementioned legislative acts has accelerated innovation for large-scale "CO_2 only" RC&H solutions, giving rise to technologies aimed at achieving this target. Among these, the ever-growing uptake of commercial "CO_2 only" around the world can be ascribable to the conception of the multi-ejector concept [8], conceived by Hafner et al. [9,10]. Also, its adoption is supposed to offer considerable energy conservations in other high energy-demanding buildings (e.g., hotels, gyms, spas) too, even in warm weathers.

In this investigation, the state-of-the-art multi-ejector based solutions for supermarkets and their technological aspects are presented. Furthermore, the potential energy benefit as well as some relevant field/laboratory data are shown. Also, the findings associated with the implementation of the multi-ejector concept in other applications are described. At last, the remaining challenges requiring to be faced are summarized.

2. Multi-Ejector Concept

The performance of basic "CO_2 only" vapor-compression systems is significantly more sensitive to cooling medium temperature than HFC-based units. As a consequence of the low critical temperature of R744 (i.e., about 31 °C), in fact, transcritical running modes can commonly take place. These cause a large exergy destruction during the throttling process [11], which negatively impacts on the overall efficiency of the system. On the other hand, the significant irreversibilities associated with the expansion valve induce to take into considerations expansion work recovery devices, such as expanders and two-phase ejectors. Their favorable contribution is, in fact, much more relevant than in traditional HFC applications, in which the inefficiencies related to the throttling process are more limited. In the last years, two-phase ejectors have been gaining popularity thanks to their intrinsic simplicity. In general terms, it has been shown that the implementation of a two-phase ejector leads to greater opportunity for energy efficiency improvement [12]. A simple transcritical R744 vapor-compression unit employing an ejector aimed at expansion work recovery is sketched in Figure 1. The refrigerant coming out of the gas cooler/condenser (thermodynamic state 3, identifying the high pressure) is expanded and accelerated through the motive nozzle (thermodynamic state 4). Due to the pressure difference between the expanded refrigerant and the working fluid exiting the evaporator (thermodynamic state 10, identifying the low pressure), the low pressure stream is entrained into the suction nozzle (thermodynamic state 5). Both streams are then mixed in the mixing chamber (thermodynamic state 6) and a part of the remaining kinetic energy of R744 is converted into a pressure increment via the diffuser (thermodynamic state 7, identifying the intermediate pressure). The adoption of a two-phase ejector in place of an expansion valve permits benefiting from two main energy advantages: (i) rise in refrigerating effect as the refrigerant enters the evaporator at lower vapor quality and enthalpy; (ii) decrease in compressor power input since the refrigerant is pre-compressed by the ejector from the evaporator pressure to the intermediate one (IP).

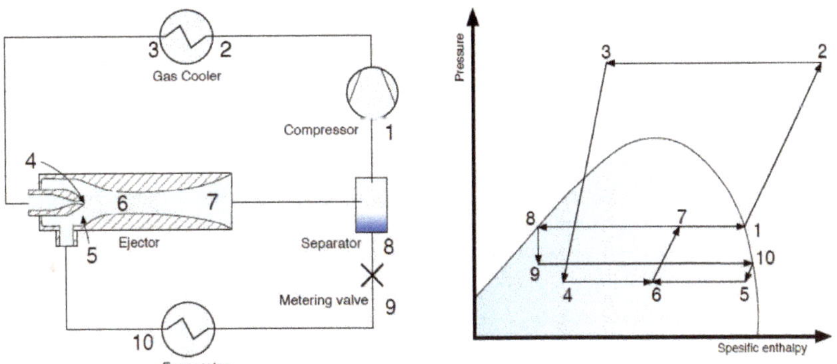

Figure 1. Schematic of a simple transcritical R744 vapor-compression system equipped with a two-phase ejector for expansion work recovery (left-hand side) and its p-h diagram (right-hand side) [13].

The performance of an ejector for expansion work recovery is commonly described with the aid of 4 metrics, i.e., mass entrainment ratio (ω), suction pressure ratio (Π), pressure lift (P_{lift}) and expansion work recovery efficiency, being generally indicated as ejector efficiency ($\eta_{ejector}$). The mass entrainment ratio (Equation (1)) refers to the ratio of the suction mass flow rate to the motive mass flow rate and evaluates the ability of the ejector to entrain (or pump) the refrigerant:

$$\omega = \frac{\dot{m}_{suction\ nozzle}}{\dot{m}_{motive\ nozzle}} \tag{1}$$

The suction pressure ratio (Equation (2)) and the pressure lift (Equation (3)) assess the ratio of the ejector outlet pressure to the ejector suction pressure and difference between the ejector outlet pressure and the ejector suction pressure, respectively. These are employed for quantifying the lift that the ejector can provide to the working fluid.

$$\Pi = \frac{P_{diffuser\_outlet}}{P_{suction\ nozzle\_inlet}} \tag{2}$$

$$P_{lift} = P_{diffuser\_outlet} - P_{suction\ nozzle\_inlet} \tag{3}$$

The ejector efficiency (Equation (4)) defines the actual amount of work recovered by the ejector with respect to the overall work recovery potential. As suggested in [14], this can be computed as the power used for compressing the suction stream isentropically from suction inlet to diffuser outlet pressure divided by the theoretical maximum amount, which could be recovered via an isentropic expansion of the motive stream from motive inlet to diffuser outlet pressure.

$$\eta_{ejector} = \frac{\dot{W}_{recovered}}{\dot{W}_{recoverable\_max}} = \omega \cdot \frac{h\left(P_{diffuser\_outlet}, s_{suction\ nozzle\_inlet}\right) - h_{suction\ nozzle\_inlet}}{h_{motive\ nozzle\_inlet} - h\left(P_{diffuser\_outlet}, s_{motive\ nozzle\_inlet}\right)} \tag{4}$$

It was found that the efficiency of R744 ejectors available in the literature is usually between 0.2 and 0.3, whereas the efficiency associated with both R410A and R134a ejectors is generally below 0.2 [12].

A solitary constant-geometry ejector cannot ensure an effective control of the heat rejection pressure and, simultaneously, implement expansion work recovery accurately [15]. In order to overcome such a drawback, the multi-ejector concept was formulated. As schematized in Figure 2, this relies on a block hosting several fixed geometry ejector cartridges of various size and arranged in parallel. The multi-ejector modules available on the market feature of 4–6 vapor ejectors and, as regards food retail applications, 2 liquid ejectors. At least one of the vapor ejectors is in operation and the necessary capacity is constantly fulfilled by changing their combination, besides guarantying the occurrence of the optimal high-side working conditions in any operating mode. Therefore, the mass flow rate required to meet the cooling load is available in any running mode, permitting to handle the variable demands appropriately. The scenario involving the use of 3 out of 6 ejectors is represented in Figure 2. Solenoid valves employed for activating the ejectors are located on the upper part of the block (Figure 2), while the pressure level can be measured by using the pressure sensors (located on the right-hand side in Figure 2) for each port (i.e., high pressure side, suction and discharge ports, located from the top to the bottom on the left-hand side in Figure 2). Typically, an expansion valve is arranged in parallel with the multi-ejector block to unceasingly and effectively control the gas cooler pressure to the detriment of some of the available expansion work recovery.

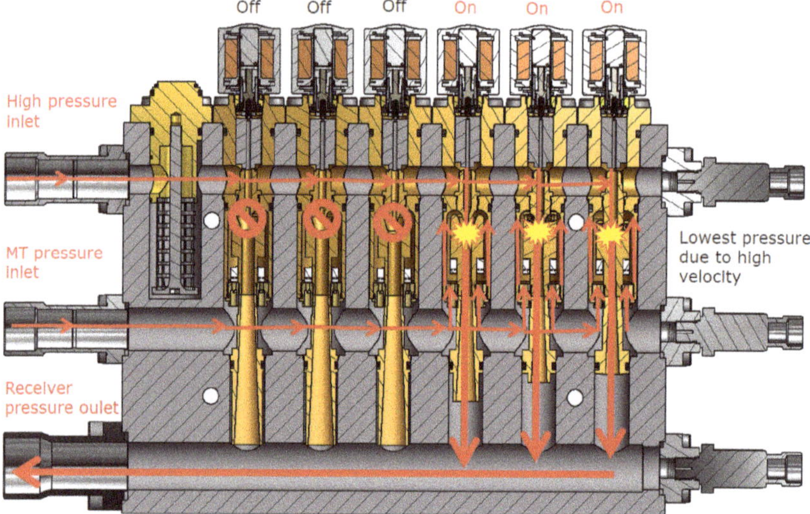

Figure 2. Sketch of the multi-ejector block [16].

As for supermarket refrigeration units, additional energy savings are obtained by overfeeding the evaporators. As a consequence, these heat exchangers can operate at a higher working temperature than conventional dry-expansion evaporators. Ejectors have been successfully implemented as a simple way to recirculate liquid out of the evaporators. It is worth remarking that evaporator overfeeding, in which excess liquid recirculation is provided by a liquid ejector with an efficiency of 8%, improves the overall annual performance by 15%, whereas a vapor ejector with a peak efficiency of 30% would lead to an annual performance enhancement by 5%, depending on the outdoor temperature profile [17].

3. Supermarket Applications

The multi-ejector concept was firstly applied to commercial refrigeration systems due to their major negative environmental impact. This is mainly caused by their significant high electricity consumptions as well as their relevant refrigerant charge losses. As an example, it was estimated that a typical American food retail store having a sale area of about 3700÷5600 m^2 consumes approximately 2÷3 GWh of electricity yearly [18]. Furthermore, the average annual leakage rate of the working fluid is roughly between 3% and 22% of the total charge [19] and R404A (GWP$_{100 \text{ years}}$ =

$3943\ \text{kg}_{CO_{2,equ}} \cdot \text{kg}^{-1}_{refrigerant}$ according to AR5) is the most widely employed refrigerant in the European food retail industry [19].

3.1. Evolution of System Layout

In the last 10 years "CO$_2$ only" supermarket refrigeration system layouts have experienced a considerable evolution, leading these solutions to move from the 1st to the 3rd generation. This significant technological development has targeted an enhancement in their energy efficiency in any climate context, with particular emphasis on the units located in warm locations.

The term "1st generation" refers to the basic R744 booster supermarket refrigeration plant layout including the flash gas by-pass valve (schematic on the left-hand side in Figure 3) [20,21]. Currently such a solution presents on the high pressure (HP) side one or two de-superheater(s) devoted to heat recovery for space heating and DHW purposes and located upstream of the air-cooled gas cooler/condenser [22]. Therefore, R744 coming out of the latter heat exchanger is throttled to the IP, leading to the generation of a liquid/vapor mixture. The liquid fraction, being separated in the IP liquid receiver, is employed for feeding the medium (MT) and low temperature (LT) evaporators. The refrigerant exiting the LT evaporators is then drawn by the "booster" (i.e., LT) compressors and compressed to the medium pressure (MP). The pressure of the vapor in the IP liquid receiver is reduced to MP via flash gas by-pass valve, which is then mixed with the refrigerant coming out of both the MT evaporators and LT compressors, before being compressed by the MT compressors to HP. Many investigations on this solution (e.g., [5,23,24]) are currently available in the open literature. These systems are very popular in Northern Europe (i.e., cold areas), as they perform similarly to or better than a conventional HFC-based unit at outdoor temperatures up to about 24 °C [25].

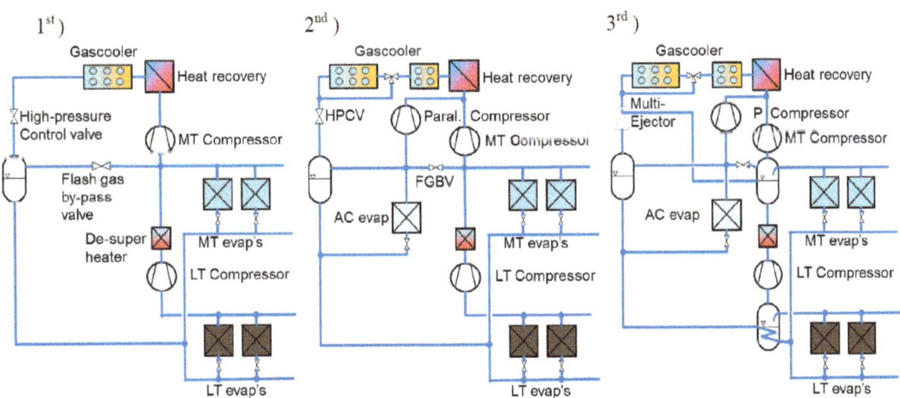

Figure 3. Schematic of the 1st, 2nd and 3rd generation of "CO$_2$ only" booster supermarket refrigeration system layouts [20].

A first step towards the adoption of such HFC-free solutions in warm areas is offered by the 2nd generation [20,21], i.e., parallel compression-equipped units (schematic on the middle in Figure 3). The adoption of parallel (or auxiliary) compression permits MT compressors to be unloaded for the sake of use of auxiliary compressors, which perform at a higher suction pressure (and thus at more favorable operating conditions) than the former. As a consequence, energy savings can be achieved compared to the 1st generation. The investigation by Karampour and Sawalha [26], based on some filed measurements, recently revealed that parallel compression offers high energy savings in cold climates, while it is not fully suitable for hot climates. Parallel compression-based R744 systems have been widely studied by many researchers (e.g., [27–29]).

The proliferation of transcritical R744 refrigeration plants in the commercial sector is thus expected to occur with the aid of the concurrent implementation of several energy efficient measures [8], such as parallel compression, overfed evaporators [30,31] and multi-ejector concept, i.e., via the 3rd generation of "CO$_2$ only" supermarket refrigerating systems (schematic on the right-hand side in Figure 3) [20,21]. Therefore, the expansion work is partially recovered and used by the vapor ejectors for moving part of the refrigerant after MT evaporators to the parallel compressors, which operate at a higher suction pressures and thus leading to considerable energy savings, especially at severly high outdoor temperatures.

The unique properties of CO$_2$ favor the effective recover of heat for space heating and DHW purposes [22,32]. This permits these HFC-free solutions to additionally increase their energy saving [22,33] and reduce their carbon footprint [34], besides offering satisfactory payback times [35]. The waste heat utilisation process, which has become an integral part of any "CO$_2$ only" supermarket refrigeration plant, further promotes the use of a multi-ejector block as transcritical operating conditions commonly take place in heating mode [22].

3.2. Evolution of Multi-Ejector Equipped System Layout

The first proposed R744 multi-ejector enhanced parallel compression system architecture, allowing the overfeeding of MT evaporators, is sketched in Figure 4. Additionally, Minetto et al. [36] recommended the adoption of an internal heat exchanger (IHX) as presented in Figure 5 to overfeed the LT evaporators too. The additional energy benefits derived from adopting such a measure has led the system layout (and similar system architectures) to become the preferred solutions on the part of the end-users as multi-ejector based units are chosen. However, as the LT evaporators are overfed, the presence of a low pressure (LP) accumulator to trap the liquid before compressors is mandatory, since liquid cannot possibly evaporate in the heat exchanger indicated as IHX D in Figure 5 [37]. Furthermore, IHX C (Figure 5) aims at heating up the refrigerant before this is drawn by the LS compressors in any operating conditions.

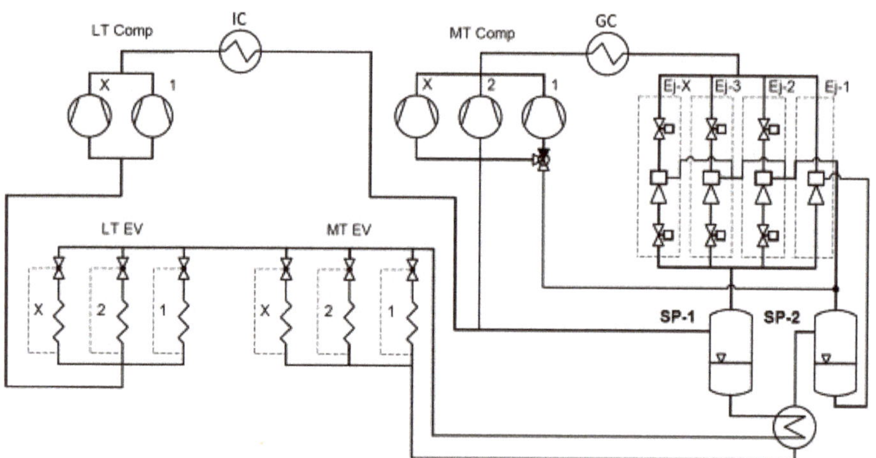

Figure 4. Schematic of a transcritical R744 booster supermarket refrigeration system outfitted with multi-ejector module and having MT overfed evaporators [10].

Fully-integrated (or all-in-one) solutions are tailor-made units, which successfully provide the entire refrigeration, air conditioning (AC) and DHW reclaims, alongside satisfying most of or even the whole space heating load of the selected supermarket [21,26,38,39]. According to [6], all-in-one transcritical CO$_2$ supermarket refrigeration plants equipped with multi-ejector block (e.g., schematic on the right-hand side in Figure 3) are thought to significantly bring the total investment, maintenance and running costs down, besides offering other advantages [26], such as compactness, reduction in complexity of communication between all the entities responsible for operating the various units. Also, the adoption of these units permits overcoming the persisting problem represented by the selection of the best refrigerant for AC and heating applications. The system layout schematized in Figure 6 and adopted in a supermarket in the North of Italy simultaneously implements both the all-in-one and multi-ejector concept [21].

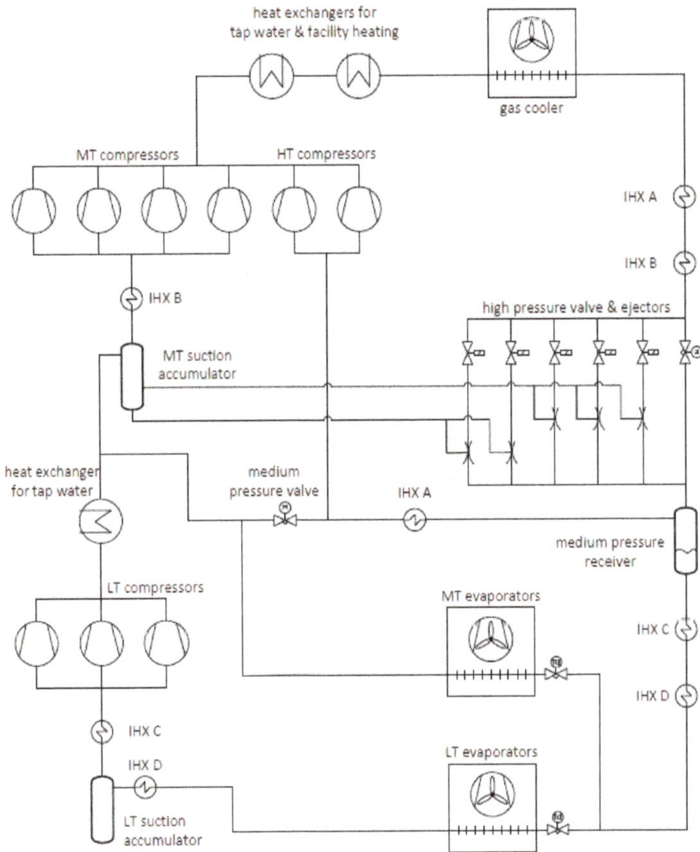

Figure 5. Schematic of a transcritical R744 booster supermarket refrigeration system outfitted with multi-ejector module and having both MT and LT overfed evaporators [37].

Hafner [39] predicted that next generation of commercial "CO$_2$ only" refrigerating systems will be characterized by both the application of the all-in-one concept and the use of two multi-ejector modules (i.e., one dedicated to refrigeration loads and the other to AC purposes).

The unit architecture sketched in Figure 7 represents a suitable solution for supermarkets located in Southern Europe or Middle East [40]. The solution features: (1) the presence of two multi-ejectors modules (i.e., one dedicated to refrigeration loads and the other to AC purposes); (2) the implementation of the "principle of pivoting" (see Section 3.4.1); (3) the use of MT and LT overfed evaporators; (4) an exterior heat exchanger (HX) operating as an additional evaporator to increase the amount of recoverable heat in wintertime and as a gas cooler in AC mode; (5) an auxiliary heat sink upstream of the multi-ejector blocks to cool R744 down and (6) the integration of ice-water cooling evaporators coupled with the AC multi-ejector block to increase the suction pressure of the parallel compressors.

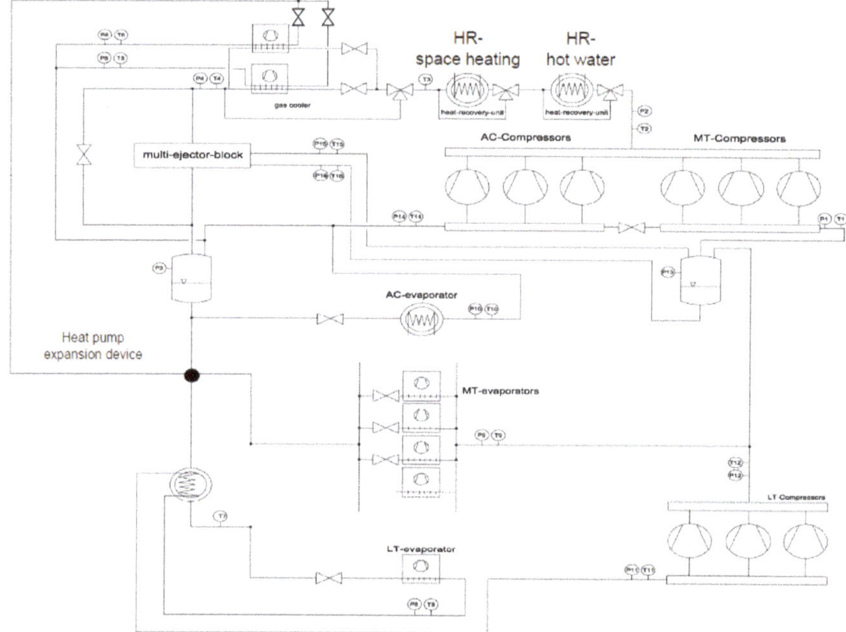

Figure 6. Fully-integrated R744 multi-ejector enhanced parallel compression system installed in a food retail store located in Spiazzo (Northern Italy) [21].

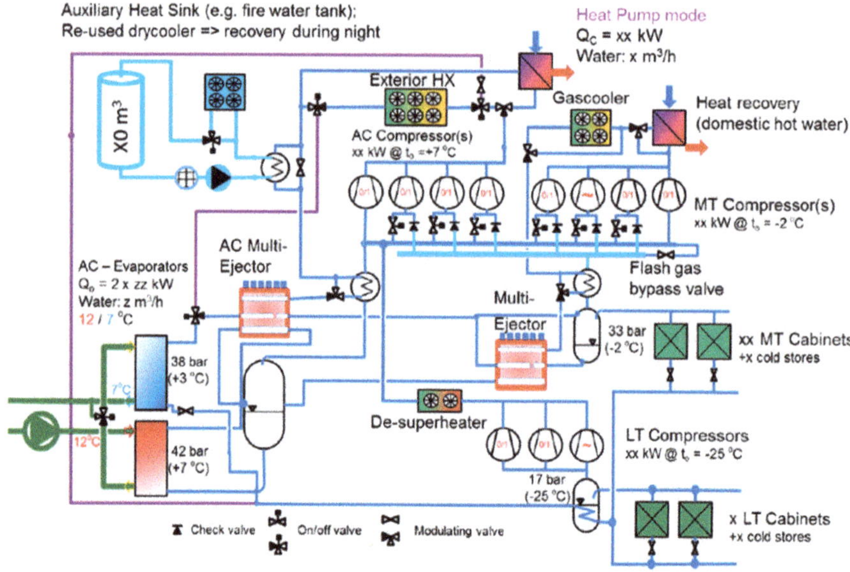

Figure 7. Fully-integrated R744 multi-ejector enhanced parallel compression system suitable for supermarkets located in warm areas [40].

The adoption of a secondary fluid between the refrigeration plant and the AC/heating unit introduces additional penalizations [40,41]. These can be avoided thanks to the favorable thermo-physical properties of R744, which can be employed in direct cooling and heating fan coils and air curtains installed inside the building. The implementation of such a technique would lead to many benefits, such as higher energy efficiency, reduction in number of components, possible decrease in investment costs, no corrosiveness issues. A refrigeration plant layout based on this concept is presented in Figure 8. In AC mode, the expansion valves upstream of the fan coils and air curtains, operating as evaporators, guarantee the correct amount of R744 to each unit. As the heating mode takes place, the gas cooler is by-passed and thus the heat is directly transferred into the building by the fan coils and air curtains.

The solution presented in Figure 9 enables recovering part of the available expansion work as the AC operations take place as well. The appropriate amount of CO_2 in the fan coil or air curtain is obtained via the modulating 3-way-valve located downstream of the ejectors. In heating mode, the gas cooler is by-passed and thus the heat is rejected directly into the building by the unit (fan coil or air curtain).

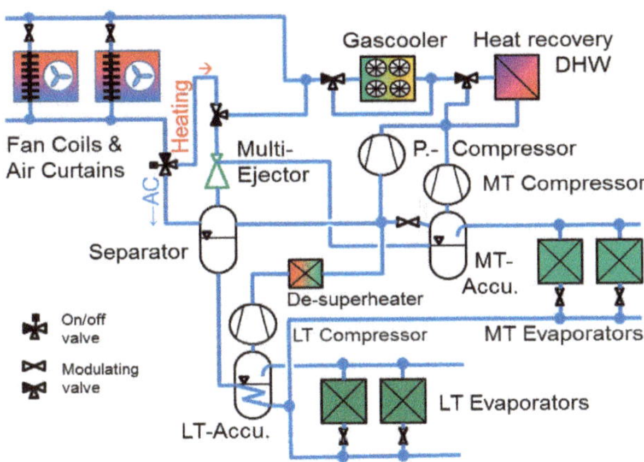

Figure 8. Integration of direct heating and cooling fan coils and air curtains in a transcritical R744 supermarket refrigeration system (multi-ejector block partly by-passed in AC mode) [40].

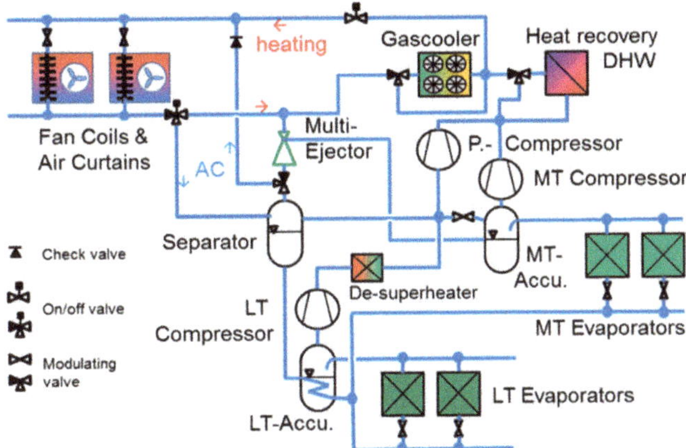

Figure 9. Integration of direct heating and cooling fan coils and air curtains in a transcritical R744 supermarket refrigeration system (ejectors employed in AC mode as well) [40].

3.3. Multi-Ejector Based Solutions without Integration with Air Conditioning Unit

3.3.1. Technological Aspects

The use of R744 ejectors requires technological assessments under multiple aspects. The first challenge is related to the ability of the multi-ejector module to provide HP control and optimization under variable operating conditions.

The experimental results presented in [15] suggested that the heat rejection pressure can be satisfactorily controlled by a multi-ejector block in commercial refrigeration applications. As showed in Figure 10, in fact, the researchers evaluated similar profiles of the discharge pressure control error caused by a rapid change in both load and outdoor temperature between a standard high pressure electronic expansion valve (HVP) and a multi-ejector module.

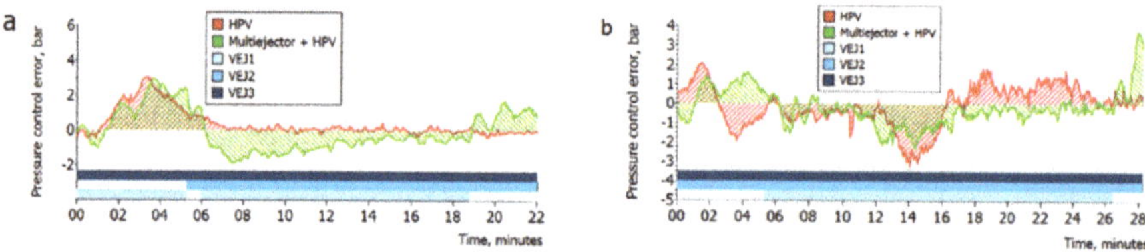

Figure 10. Deviation between the actual value and set-point value for the heat rejection related to a rapid change in load (**a**) and in outdoor temperature (**b**) [15].

The field measurements gathered in [42,43] revealed that a multi-ejector module can successfully switch the gas cooler/condenser from floating condensing to heat recovery mode and vice versa (as showed in Figure 11 after 13:00) as well as the MT evaporators from superheated to overfed mode and vice versa. In the reference example, the authors also found that the minimum number of ejectors needed to reach the discharge pressure set points without too many on/off switches of these devices was 3. Finally, they demonstrated that the control system can suitably handle possible blockages of an ejector nozzle.

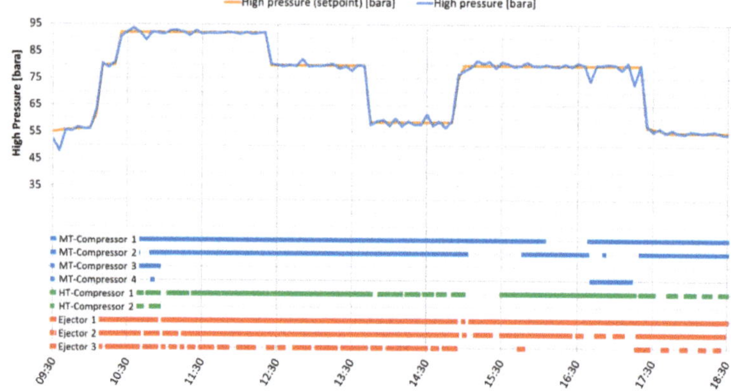

Figure 11. Operating conditions aimed at demonstrating the successful switching from floating condensing to heat recovery mode and vice versa via multi-ejector concept [42].

Also the oil management can be properly performed, taking care of recovery both at compressor discharge and from the liquid separator [15,21].

The significant use of the parallel compressors [21,37] allows reductions in their maintenance issues all over the year compared to the conventional booster configuration [44].

According to [45], the use of the multi-ejector module can successfully bring down the installed displacement of compressors. Furthermore, [17] highlighted that the implementation of the multi-ejector concept allows:

- Significantly decreasing the compressor discharge temperature. This is shown in Figure 12, in which curve refers to the typical operating conditions of an overfed evaporator, whereas curve b and c are related to two conventional running modes of dry-expansion evaporators. Therefore, these results highlight considerable benefits to the lifetime of the lubricant, components on the discharge line and de-superheater for heat recovery;
- Improved protection against liquid in the compressor suction manifolds thanks to both the adopted active methods with the purpose of limiting the liquid level and the MP liquid receiver;
- A reduction in total installed swept volume in relation to a single compression system;
- An enhanced overall energy efficiency at outdoor temperatures up to between 40 °C and 42 °C.

Curve	Suction pressure Saturated press. [°C]	Suction temperature [°C]
A	-3	+2
B	-8	20
C	-8	25

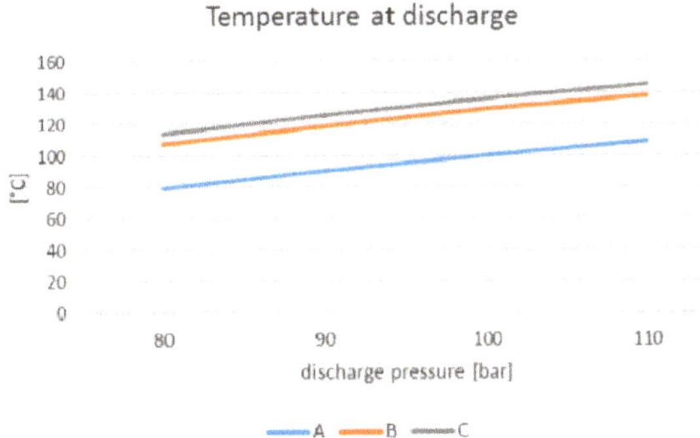

Figure 12. Effect of the heat rejection pressure on the compressor discharge temperature for the 3 investigated scenarios [17].

The experimental study in [46] demonstrated that a multi-ejector block leads to efficient and stable performance over the whole range of the investigated operating regime for commercial refrigeration applications.

3.3.2. Theoretical Assessments/Statements

Hafner et al. [9] estimated that a R744 multi-ejector enhanced parallel compression solution (same as that in Figure 4) in a food retail store located in Mediterranean Europe enables an energy saving by approximately 11% over a conventional booster system.

The dynamic simulations implemented by Hafner at al. [10] (same solution as that in Figure 4) showed that typical Coefficient of Performance (*COP*) increments in the refrigeration mode by 17% in Athens (Greece), 16% in Frankfurt (Germany) and 5% in Trondheim (Norway) can be achieved in summertime, while the ones associated with wintertime were found to be from 20% to 30% over a conventional booster plant.

The results obtained by Minetto et al. [44] brought to light that the adoption of the multi-ejector concept leads to an energy reduction by 22.5% compared to a basic "CO_2 only" unit for a MT commercial refrigeration application operating in Bari (Southern Italy).

Pisano [47] stated that a food retail store located in Bari can decrease its energy consumption by almost 30% by replacing a conventional booster unit with a multi-ejector based booster solution (similar to that in Figure 4) (Figure 13).

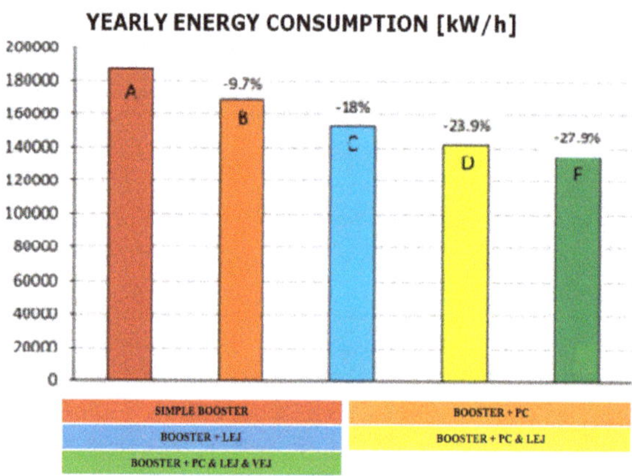

Figure 13. Energy savings of various transcritical R744 booster supermarket refrigeration systems compared to a conventional transcritical R744 booster supermarket refrigeration plant in Southern Italy [47].

Schönenberger [37] claimed that, compared to the solution with parallel compression, an energy saving between 15% and 25% can be accomplished with the aid of the multi-ejector concept (same system as that in Figure 5), depending on the heat demand, application and climate conditions.

The assessment by Gullo et al. [48] revealed that R744 ejector supported parallel vapor compression systems (similar to those in Figures 4 and 5) reduce the energy consumption from roughly 20% to 27% in comparison with a R404A unit in an average-size food retail store located in Mediterranean Europe. At the same boundary conditions, parallel compression offers, at best, an energy conservation by 6.4%.

Gullo et al. [49] estimated that R744 multi-ejector enhanced parallel compression units (similar to those in Figures 4 and 5) offer energy savings between 17.8% (in Oslo, Norway) and 26.7% (in Athens) in relation to a conventional booster system and between 16.9% (in London, UK) and 23.4% (in Oslo) over a parallel compression-based solution. In comparison with a R404A unit, the adoption of the multi-ejector concept leads to a decrease in annual electricity intake from 24.6% (in Athens) to 37.1% (in Oslo).

Due to the EU F-Gas Regulation 517/2014 the adoption of R404A-based supermarket systems will not be allowed after 2022. Consequently, cascade/indirect loop arrangements are drawing interest, especially in warm locations. However, these are expected to be less efficient, as suggested in a recent report by the European Commission [50] and recently confirmed in [51,52].

Gullo and Hafner [51] carried out an investigation based on the operating conditions of a typical supermarket and several American cities, including cold, moderate and warm weathers. The results obtained showed that R744 multi-ejector enhanced parallel compression systems (similar to those in Figures 4 and 5) offer energy conservations from 17.3% to 37.8% compared to a R404A-based unit. Furthermore, energy savings up to 26% were estimated for these solutions in warm locations. At best, R1234ze(E)/R744 indirect arrangements consumed 10.5% less electricity over the R404A-based unit at the same boundary conditions.

The results of the study by Gullo et al. [52], which was based on an average-size supermarket located in several cities positioned below the so-called "CO$_2$ equator" (average yearly temperature between 14.1 °C and 18.9 °C), suggested that compared to R404A direct expansion units:

- Multi-ejector based solutions (similar to those in Figures 4 and 5) can reduce the energy consumption from 18.6% to 28.6%;
- The r1234ze(E)/R744 indirect arrangement with MT and LT flooded evaporators and the r134a/R744 cascade solution present some modest energy savings;
- The other evaluated systems (i.e., R1234ze(E)/R744 indirect arrangement with MT flooded evaporators, R290/R744 indirect arrangement with and without LT flooded evaporators, R450A/R744 cascade solution and R513A/R744 cascade system) are not suitable candidates.

Figure 14 summarizes the outcomes of the previous investigation in terms of reduction in Total Equivalent Warming Impact (TEWI) relying on R404A direct expansion units as the baseline. The authors [52] found that the adoption of the multi-ejector concept allows reducing the environmental impact from 50.7% to 90.6%, while the aforementioned R1234ze(E)/R744 indirect unit and the R134a/R744 cascade arrangement decrease the carbon footprint from 39.1% to 87.7% and from 28.8% to 65.8% TEWI, respectively.

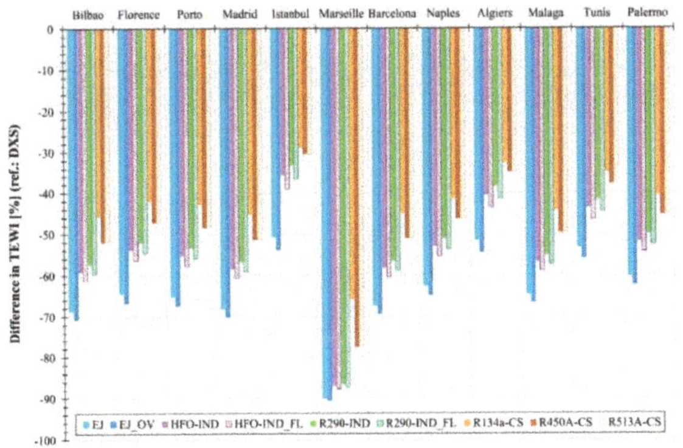

Figure 14. Difference in TEWI among various supermarket refrigeration solutions (DXS: R404A unit; EJ: multi-ejector based system with MT overfed evaporators; EJ_OV: multi-ejector based system with MT and LT overfed evaporators; -IND: indirect arrangements with MT flooded evaporators; -IND_FL: indirect arrangements with MT and LT flooded evaporators; -CS: cascade arrangements) in several locations positioned below the CO_2 equator ($t_{MT} = -4/-10\ °C$, $t_{LT} = -27/-35\ °C$, $\dot{Q}_{MT} = 120$ kW, $\dot{Q}_{LT} = 25$ kW) [52].

Madsen and Kriezi [53] recently showed that the implementation of the multi-ejector concept (solution similar to that in Figure 4) leads to energy savings from 13% to 29% compared to the conventional booster unit in locations featuring an average annual temperature between 0 °C and 30 °C (Figure 15).

| | | Energy saving compared to Booster CO2 system | | | | | |
| | | HP ejectors | LP ejectors | Flooded Operation | | | |
Average Annual temp [C]	Parallel compression	Parallel comp. + HP ejectors	Booster sys + LP Ejector	Booster sys. with Liquid ejector	Parallel comp. sys . with liquid ejector	LP Ejector system with Liquid ejectors	HP ejector system with Liquid ejectors
0	5%	6%	2%	10%	15%	13%	17%
5	5%	7%	3%	10%	16%	14%	18%
10	6%	9%	5%	10%	16%	16%	20%
15	7%	12%	9%	10%	17%	19%	23%
20	7%	15%	13%	10%	18%	24%	27%
25	8%	17%	15%	10%	18%	26%	29%
30	6%	16%	15%	10%	17%	27%	27%

Figure 15. Energy saving of various transcritical R744 booster supermarket refrigeration systems at different average annual temperatures compared to a conventional transcritical R744 booster supermarket unit [53].

Also, LP and HP in Figure 15 respectively refer to vapor ejectors with low and high pressure lift. At the same boundary conditions, parallel compression offers reductions in energy consumption from 5% to 8%.

Gullo et al. [54] concluded that the adoption of multi-ejector based solutions is also suggested as a result of the application of the advanced exergy analysis. Such an assessment is widely recognized as the most powerful thermodynamic tool to suitably assess the performance of any energy system. In addition, the researchers found that the implementation of the multi-ejector concept leads a conventional booster unit to a decrease by about 39% in total irreversibilities at the outdoor temperature of 40 °C. This outcome was a consequence of: (i) a reduction by approximately 36% in avoidable irreversibilities taking place in the gas cooler; (ii) a decrement by about 39% in avoidable inefficiencies occurring in the compressors discharging to the heat rejection pressure; (iii) a halving of the avoidable exergy destruction related to the MT evaporators; (iv) a decrease by about 40% in total inefficiencies associated with the main expansion device.

3.3.3. Laboratory and Field Experimental Assessments

Laboratory Experimental Assessments

The experimental assessment fulfilled in [13] revealed that efficiencies of an individual ejector hosted in a multi-ejector block up to 0.3 can be measured with respect to the heat rejection pressure and temperature, the pressure lift and the evaporation pressure.

The in-depth experimental investigation in [15] brought to light that the ejector efficiencies above 0.3 can be accomplished over a broad range of the investigated operation conditions. Furthermore, the results obtained revealed that the assessed ejector efficiencies were higher than those previously gathered. However, the overall multi-ejector efficiency is gradually penalized as the expanded mass flow rate grows owing to the increasing stream irreversibilities (e.g., imperfect mixing of individual flows exiting the ejectors), although the recorded values of efficiency were found to be above 0.2. The researchers mapped the motive nozzle mass flow rate depending on the inlet density and inlet pressure (Figure 16a). In addition, it was seen that the optimum running modes of an ejector having a given geometry can be attained by varying the suction pressure ratio with respect to the gas cooler outlet conditions (Figure 16b). Furthermore, [15] discovered that the compressor efficiency has a considerable influence on the overall system efficiency with respect to the selected combination of ejector cartridges. Additionally, a maximum *COP* enhancement of 9.8% was evaluated as only 50% of the total mass flow rate went through the ejectors and the compressor was operating at the highest efficiency.

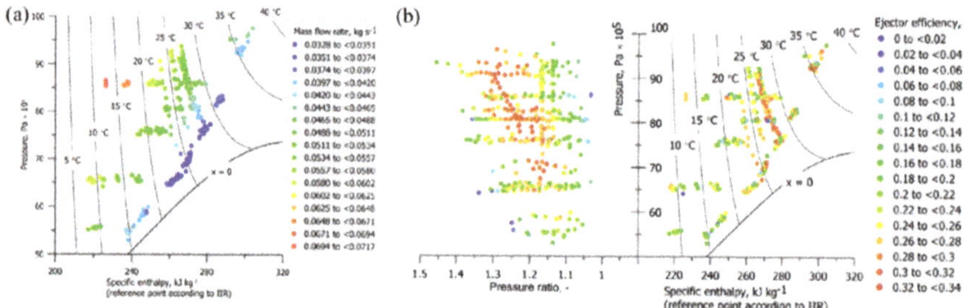

Figure 16. (a) Motive nozzle mass flow rate as a function of the motive nozzle inlet conditions for one of the investigated ejectors; (b) Ejector efficiency as a function of the motive nozzle inlet conditions and suction pressure ratios for one of the investigated ejectors [15].

Fredslund et al. [55] observed that vapor ejector efficiencies measured in the laboratory (Figure 17a) are comparable (and above 0.25) to those estimated from real installations operating at the typical operating conditions (i.e., P_{lift} = 6 bar) (Figure 17b). According to the authors, compressor sizes (so as to satisfy the required capacity and have the parallel compressors as long as in operation), pressure lift and the oil return design should be carefully considered.

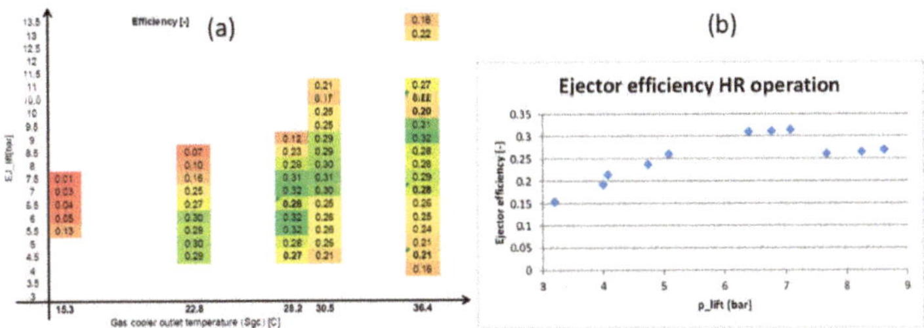

Figure 17. Comparison between the ejector efficiencies measured in the laboratory (**a**) and those estimated in a real application performing in heat recovery mode (**b**) [55].

Haida et al. [56] proved that the adoption of the multi-ejector block permits increasing *COP* and exergy efficiency up to 7% and 13.7% over the parallel compression-based unit, respectively. In addition, efficiencies of the multi-ejector module up to 0.33 were estimated with respect to the pressure lift and the motive and suction conditions. The researchers also pointed out that it is required to mutually taken into account the multi ejector block and the compressor rack interactions when designing the refrigeration plant.

The experimental results obtained by Pardiñas et al. [57] demonstrated that efficient MT compressors in their range of operation are compulsory to accomplish significant energy conservations (even with efficient vapor ejectors) as vapor ejectors boost parallel compressors and unload MT compressors. Thus, maintaining high compression efficiency for the MT machines even in part-load becomes a crucial factor for maximizing the system efficiency gains thanks to the ejector use.

Field Experimental Assessments

The incorporation of the multi-ejector module into a R744 booster unit enables increasing the suction pressure of the auxiliary compressors by 3÷10 bar over that of the MT compressors [43,58].

Kriezi et al. [59] suggested employing a liquid ejector designed for summer operating conditions and another for winter running modes owing to the highly fluctuating demands of food retail applications. Ejectors aimed at vapor removal can adequately pump some liquid in summertime.

The field data presented in [36] (Figure 18) demonstrated that an increment in MT by 6 K compared to a dry-expansion evaporator can be attained with the aid of the multi-ejector concept. The increase in LT by 8 K presented in Figure 18 was a result of the adoption of an IHX, as mentioned above. Both achievements can be maintained all over the year, leading to reductions in frost formation and number of defrost cycles in relation to conventional technology [17,37,42,43].

Kriezi et al. [60] estimated the energy benefits associated with the adoption of MT overfed evaporators on the part of a multi-ejector based system (similar to that in Figure 4) with the aid of field measurements collected between June and August 2017. The authors estimated that MT can be increased on average by 4.5 K, as dry-expansion evaporators with a minimum degree of superheating of 6 K perform in overfed mode, with a minimum degree of superheating set to 1 K. As a consequence, the total *COP* values can be enhanced on average by about 5% over the investigated range of running modes (Figure 19a), leading to a potential decrease in power input of approximately 4.5% (Figure 19b).

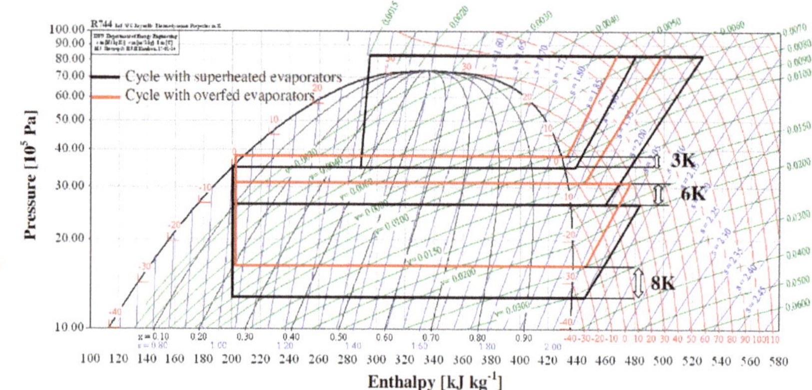

Figure 18. Comparison between the thermodynamic cycle with superheated evaporators and that with overfed evaporators [17].

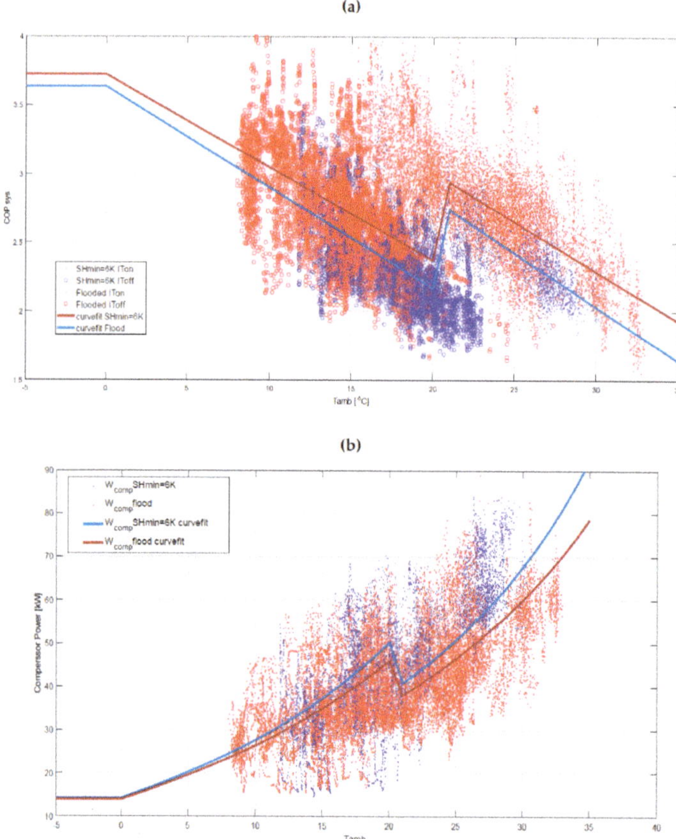

Figure 19. Comparison of total *COP* values (**a**) and total compressor power input (kW) (**b**) between the investigated system using MT dry-expansion evaporators with a minimum degree of superheating of 6 K (blue line) and the same system relying on MT flooded evaporators with a minimum degree of superheating set to 1 K (red line) [60].

The *COP* increase and the compressor power reduction were lower than expected, as the system had a high LT load (equivalent to MT capacity) and the LT evaporators were running in dry-expansion mode. However, the authors also highlighted that the energy advantages related to the parallel compressor were underestimated as well as the ones associated with the potential reduction in number of defrost cycles were not considered.

Schönenberger et al. [42] and Hafner et al. [43] demonstrated that energy savings by 10% in comparison with the solution using parallel compression and by 18% over a conventional booster system can be obtained in a supermarket (system similar to that in Figure 4) located in the region of Fribourg (Switzerland) in wintertime, respectively.

The energy consumption of the first installation relying on the multi-ejector concept (similar to that in Figure 4) is compared to that of three similar units employing parallel compression in Figure 20. It was found to offer an energy conservation by 14% over the same period of time in the Swiss climate context [42].

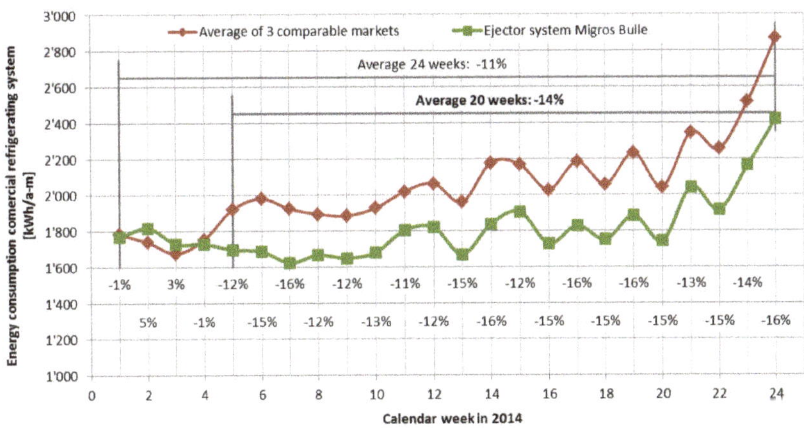

Figure 20. Energy consumption comparison between the first unit implementing the multi-ejector concept and three similar systems relying on parallel compression in the Swiss climate context [42].

3.3.4. Economic Assessments

As presented in Figure 21, the addition of vapor ejectors with high pressure lift to units relying on parallel compression is economically acceptable for supermarkets with cooling capacities over 75 kW and located in cities having an average annual temperature above 15 °C [60]. LP and HP in Figure 21 respectively indicate vapor ejectors with low and high pressure lift.

	Pay Back Liquid ejector [years]				Pay Back HP ejector solution [years]			Pay Back LP ejector solution [years]		
Average annual temperature [°C]	Booster + Liq VS Booster 40 kW	Booster + Liq VS Booster 75 kW	Booster + Liq VS Booster 150 kW	Booster + Liq VS Booster 300 kW	HP Ejector VS parallel 75 kW	HP Ejector VS parallel 150 kW	HP ejector VS parallel 300 kW	LP ejector VS Booster 40 kW	LP ejector VS Booster 75 kW	LP ejector VS Booster 150 kW
0	3.70	1.97	1.58	0.89	14.96	13.37	12.51	21.53	17.74	18.74
5	3.50	1.87	1.49	0.84	9.50	8.50	7.96	13.59	11.20	11.83
10	3.17	1.69	1.35	0.76	5.70	5.08	4.75	8.37	6.89	7.28
15	2.61	1.39	1.11	0.63	2.63	2.32	2.16	4.15	3.42	3.61
20	2.13	1.14	0.91	0.51	1.35	1.18	1.09	2.34	1.93	2.04
25	1.79	0.96	0.76	0.43	0.87	0.75	0.69	1.69	1.39	1.47
30	1.81	0.97	0.77	0.43	0.85	0.73	0.66	1.69	1.39	1.47

Figure 21. Payback period for ejector addition to various transcritical R744 booster supermarket refrigeration systems with respect to both the average annual temperature and the required cooling capacity [60].

Pisano [47] recently reported that a "CO$_2$ only" booster refrigeration system equipped with multi-ejector block (similar to that in Figure 4) (green line in Figure 22) offers a return of investments of 2 years (based on capital and running costs) compared to a basic "CO$_2$ only" booster unit (red line in Figure 22) in a supermarket located in Bari.

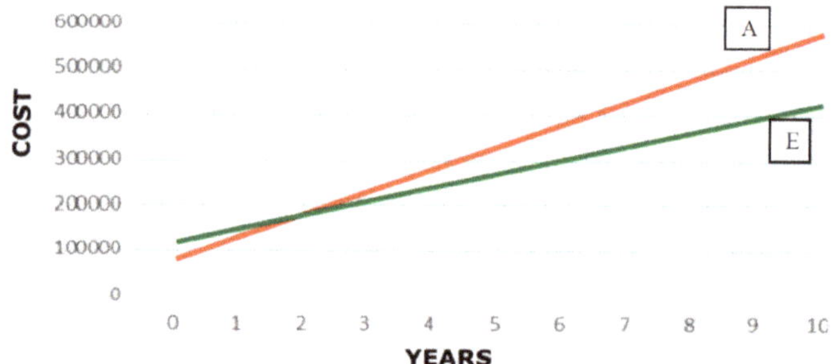

Figure 22. Comparison in terms of return on investments between a transcritical R744 booster refrigeration plant outfitted with multi-ejector block (green line) and a conventional transcritical R744 booster refrigerating unit (red line) in a supermarket located in Bari [47].

3.4. Multi-Ejector Based Solutions with Integration with Air Conditioning Unit

3.4.1. Technological Aspects

Pardiñas et al. [61] theoretically investigated the appropriate control of the heat rejection pressure as either a high ratio of AC demand to MT refrigeration load occurs or the AC ejectors operate inefficiently. The outcomes obtained suggested adopting the control technique relying on: setting the AC evaporation pressure, computing the ideal pressure lift of the AC ejectors at the selected outdoor conditions and employing it to evaluate the set point for the control of the auxiliary compressors. A decrease in this pressure lift needs to be considered as the discharge pressure control became impossible.

The adoption of the so-called "pivoting principle" for compressors was recommended in many studies [36,39,40,62]. This technique has the purpose to enable the MT and auxiliary compressors to be widely interchangeable so as to decrease the total displacement of the installed compressors. In fact, on the one hand, the total required displacement being necessary in wintertime is very low (or even zero). On the other hand, the auxiliary compressor swept volume considerably growths in summertime due to both the flash vapor generated during the expansion process and to the vapor being pre-compressed by the ejectors. The interchangeability of the MT and auxiliary compressors is implemented by linking the compressors to either the MT or the parallel suction group with the aid of on/off valves installed upstream of them (x2–x5 in Figure 23) [40], as a function of the operation conditions. Hafner [39] also claimed that this feature gives rise to a "gap-free" control of the refrigeration load, besides lowering the installation cost as well as enhancing the compactness of the unit.

According to Pardiñas et al. [62], the implementation of the "pivoting principle" favors greater system flexibility with respect to following the actual load profile (i.e., enlargement in operational range of the refrigeration plant). The researchers also highlighted that a simple and inexpensive solution, which should not influence the oil return of the compressors, to apply the aforementioned principle is potentially available. Furthermore, Pardiñas et al. [62] stated that the average annual operation time of compressors should be broadened, implying a reduction in overall ownership costs. However, the energy savings achievable through the "pivoting principle" are not noteworthy [62]. These could be increased by using the proposed pivoting technique at the discharge of the LT compressors, further promoting a growth in their number of hours in operation.

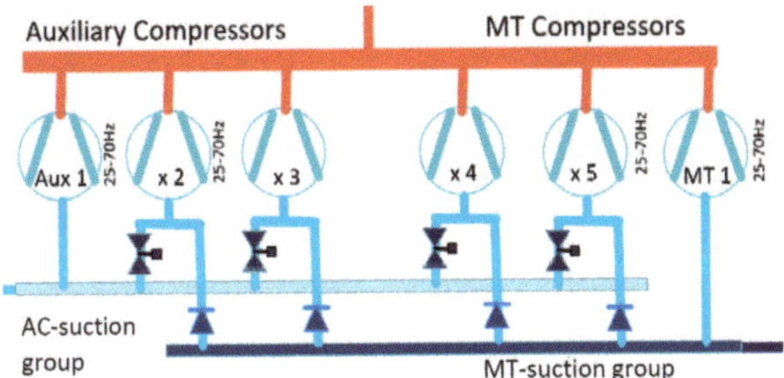

Figure 23. Schematic of the application of the "pivoting principle" to 4 out of 6 compressors [40].

3.4.2. Theoretical Assessments/Statements

Gullo et al. [48] estimated that the implementation multi-ejector concept (solutions similar to that in Figure 6) leads to energy conservations between 15.6% and 26.2% compared to conventional HFC-based units, depending on the size of both the AC equipment and the supermarket, as well as on the weather conditions.

The study by Pardiñas et al. [62] brought to light that the adoption of a (vapor) multi-ejector block is energy advantageous at outdoor temperatures above 25 °C compared to the unit relying on parallel compression (Figure 24). Further energy savings can be achieved by using an AC multi-ejector module (8.3% at 30 °C and 8.6% at 25 °C). The investigation was carried out by considering the running modes of a typical Norwegian supermarket. In addition, the authors [62] found that:

- The multi-ejector enhanced parallel compression unit with two multi-ejector blocks (one for MT load and one for AC demand) and AC evaporator located downstream of the liquid receiver is a suitable solution as high AC pressures are required;
- The multi-ejector enhanced parallel compression system with MT multi-ejector module and AC evaporator located upstream of the liquid receiver is an adequate solution as low AC pressures are necessary;
- The optimum discharge pressure is strongly related to the outdoor temperature, MT, AC evaporator position and AC evaporating pressure in transcritical operating conditions.

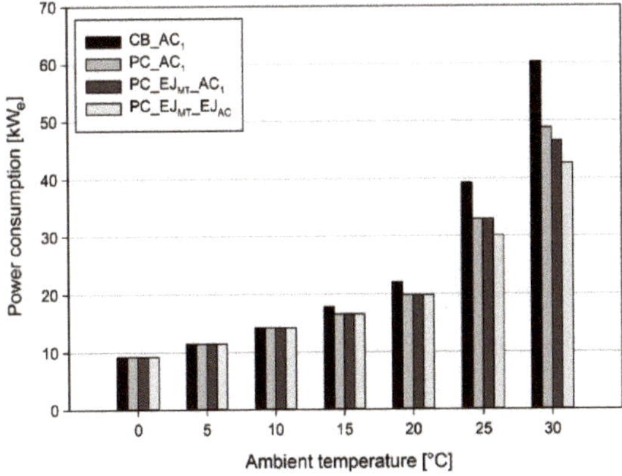

Figure 24. Electric power input of a conventional booster system (CB_AC), booster system with parallel compression (PC_AC), multi-ejector enhanced parallel compression system with (PC_EJ$_{MT}$_EJ$_{AC}$) and without (PC_EJ$_{MT}$_AC) AC multi-ejector module in a typical Norwegian food retail store (p_{MT} = 28 bar, p_{LT} = 15 bar, $p_{lift,AC}$ = 5 bar, $\dot{Q}_{MT}$ = 60 kW at $t_{outdoor}$ = 30 °C, $\dot{Q}_{LT}$ = 10 kW, $\dot{Q}_{AC}$ = 45 kW at $t_{outdoor}$ = 30 °C) [62].

Compared to R404A direct expansion units for the refrigeration loads and a R410A chiller for the AC demand, the study by Gullo et al. [52] also revealed that:

- A multi-ejector based solution integrated with the AC unit (solution similar to that in Figure 6) consumes from 19.3% to 26.9% less electricity;
- The investigated r1234ze(E)-based indirect arrangements (i.e., With and without integration with the AC equipment) offer energy savings between 4.7% and 6.4%;
- The r134a/R744 cascade system separately operating with a r1234ze(E) chiller can reduce the energy consumption from 1.9% to 4.7%;
- The other assessed solutions (i.e., R1234ze(E)-, R290-, R450A- and R513A-based systems) are not appropriate candidates.

Figure 25 summarizes the results of the aforementioned study [52] in terms of reduction in TEWI (baseline: R404A direct expansion units for refrigeration loads and a R410A chiller for AC reclaim). The researchers estimated that the implementation of the multi-ejector concept leads to a decrease in carbon footprint from 53.2% to 90.9%. Also, the aforementioned R1234ze(E)/R744 indirect solutions and R134a/R744 cascade arrangement can reduce TEWI from 40.8% to 88.5% and from 31.5% to 69.7%, respectively.

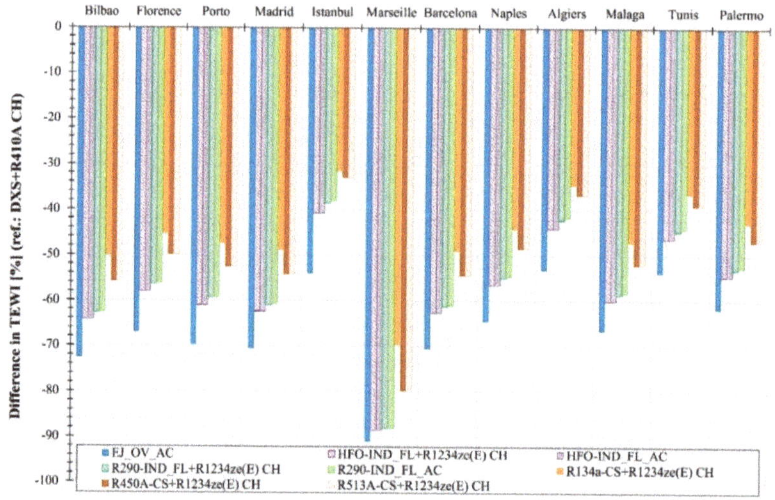

Figure 25. Difference in TEWI among various supermarket refrigeration solutions (DXS+R410A CH: R404A unit+R410A chiller; EJ_OV_AC: multi-ejector based system; -IND+R1234ze(E) CH: indirect arrangements+R1234ze(E) chiller; -CS+R1234ze(E) CH: cascade arrangements+R1234ze(E) chiller) in several locations positioned below the CO_2 equator (t_{MT} = $-4/-10$ °C, t_{LT} = $-27/-35$ °C, t_{AC} = $+3/+5$ °C, $\dot{Q}_{MT}$ = 120 kW, $\dot{Q}_{LT}$ = 25 kW, $\dot{Q}_{AC}$ = 120 kW) [52].

3.4.3. Field Experimental Assessments

The data from filed gathered by Fredslund et al. [55] in various locations showed that energy reductions from 10% to 15% at roughly 30 °C can be attained.

The field measurements collected by Hafner et al. [21] demonstrated that energy reductions between 15% and 30% over the unit relying on parallel compression can be obtained, depending on the AC load and outdoor temperature (Figure 26). Also, the researchers measured values of pressure lift from 5 bar to 10 bar. The data were gathered from a supermarket located in Spiazzo (North of Italy) between the 1st of May and the 30th of October 2015 at external temperatures ranging from 22 °C to 35 °C.

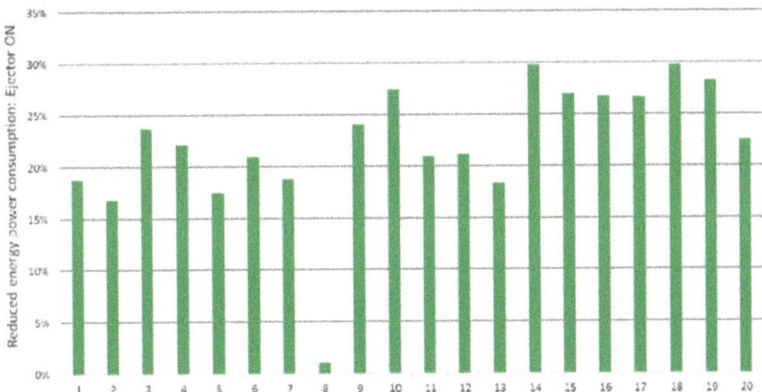

Figure 26. Reduction in energy power consumption [%] by recovering part of the available expansion work (p_{MT} = 29 bar, p_{LT} = 14 bar, $p_{lift,AC}$ = 5÷10 bar) [21].

3.5. High Ambient Temperature Countries

For climate reasons, the AC need as well as the refrigeration demand have a crucial relevance on economic, energy and environmental perspectives in high ambient temperature countries. The more the global warming effects become relevant, the more the cooling demand will increase, which will in turn accelerate climate change. According to [63], as a consequence of the enormous growth in cooling need related to emerging countries (typically hot areas), the energy reclaim for space cooling is expected to overtake space heating by 2060 and outstrip it by 60% at the end of the century. In addition, countries with emerging economies feature massive use of high-GWP refrigerants and have only recently begun the transition from hydrochlorofluorocarbons (HCFCs) to eco-friendlier working fluids, including R744.

To highlight the interest in multi-ejector based solutions in hot climates too, it is worth remarking that Middle East's first "CO$_2$ only" refrigeration plant was recently installed in a supermarket (2000 m^2, heat recapture implementation, t_{MT} = −2 °C, t_{LT} = −25 °C) in Al-Salam (Jordan). This plant was described as a test for commercial "CO$_2$ only" refrigerating units operating in high ambient temperature climate context [64].

The theoretical study by Kvalsvik et al. [65] showed that a transcritical CO$_2$ supermarket refrigeration unit presenting two multi-ejectors modules features 53% higher power input than separated R410A-based units at the design outdoor temperature of 45 °C. However, the energy benefits associated with the use of overfed evaporators were not assessed as well as the evaluation should have been performed on an annual basis.

Singh et al. [66,67] highlighted with the aid of experimental data that R744 ejector supported parallel vapor compression systems with and without integration with the AC equipment offer stable operations at high outdoor temperatures.

Blust et al. [68] experimentally showed that a multi-ejector unit is a feasible solution for the Indian climate context. The results obtained suggested that a maximum value of exergy efficiency of 0.387 can be achieved at an AC evaporator temperature of 12 °C, a MT evaporator temperature of −6 °C, a LT evaporator temperature equal to −29 °C, intermediate pressure of 52 bar and R744 gas cooler exit temperatures of 46 °C.

Singh et al. [69] implemented an experimental campaign based on an AC evaporator temperature between 7 °C and 11 °C, a MT evaporator temperature of −6 °C, a LT evaporator temperature equal to −29 °C, intermediate pressure of 44 bar and R744 gas cooler outlet temperatures between 36 °C and 46 °C. The results revealed that a maximum total *COP*, a maxim cooling *COP* and a maximum exergy efficiency equal to 4.2 (at 36 °C), 2 (at 36 °C) and 0.315 (at 46 °C) can be accomplished.

Singh et al. [67] experimentally showed that reductions in parallel compression consumption up to 10.7% can be achieved by ranging the intermediate pressure between 44 bar and 48 bar at R744 gas cooler exit temperatures between 36 °C and 46 °C. The assessment was carried out by considering an AC evaporator temperature between 7 °C and 11 °C, a MT evaporator temperature of −6 °C and a LT evaporator temperature equal to −29 °C.

The experimental assessment by Singh et al. [66] proved that the use of liquid ejectors leads to an increase in operating pressure of MT evaporators by 4.5% as well as a decrease in compressor power input by 5.5% at the R744 gas cooler exit temperature of 46 °C. The MT evaporators operated at −6 °C in dry-expansion running modes.

4. Other Applications

As mentioned above, CO_2 represents a promising working fluid for several applications. The peculiar properties of R744 make this refrigerant particularly suitable for water heating [70,71]. Cecchinato et al. [72] experimentally showed that an air-cooled CO_2 chiller for commercial refrigeration presents *COP* values between 3.1 and 2.0 at outdoor temperatures from 18 °C to 35 °C. However, reversible transcritical R744 heat pumping units are still limited to niche sectors in warm areas owing to the substantial energy efficiency penalizations taking place in AC mode [73]. The adoption of a two-phase ejector allows a R744 heat pump unit to attain the same seasonal efficiency as a R410A system in residential applications located in mild climates (i.e., north-east of Italy) [73]. Further energy advantages are supposed to be obtained with the aid of enhancement strategies similar to those implemented for supermarkets (i.e., multi-ejector concept adoption, direct heating and cooling) [10,73].

4.1. Theoretical Assessemnts

Schoenenberger and Fraga [74] theoretically estimated that a multi-ejector based "CO_2 only" vapor-compression system has an average value of *COP* 3.6% lower than that of the reference R717/R744 cascade arrangement with secondary loop. The evaluation was based on industrial application and the climate context of Valencia (Spain). The selected baseline was the most preferred ultra low-GWP alternative to the currently employed units (i.e., R507A-based solutions) for the investigated application, i.e., fish processing plant. However, the authors highlighted that the higher energy costs associated with R744 as the only refrigerant can be compensated by the reduction in maintenance costs (between 15% and 20% lower) as well as in investment cost (15% lower at the present time and 22% as a cost forecast for 2020) compared to the R717-based cascade system. As a consequence, in August 2017 two racks based on the schematic represented in Figure 27 were installed to substitute a R507A-based unit in a fish processing plant close to Valencia (Spain).

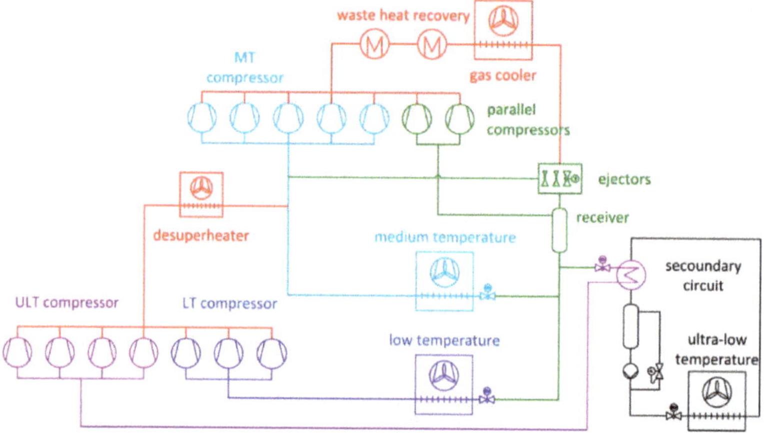

Figure 27. Schematic of the multi-ejector based "CO_2 only" vapor-compression systems installed in a fish processing plant in Valencia (t_{MT} = +2 °C, t_{LT} = −20 °C, $t_{ultra\,LT}$ = −40 °C, $\dot{Q}_{total}$ = 764 kW) [74].

The units featured evaporators being partially overfed at all the evaporating temperatures as well as heat recovery implementation devoted to produce hot water at +78 °C (heating capacity of 150 kW) for room cleaning, steam generation, etc. Another emerging application is currently related to reversible chillers for high energy-demanding buildings.

4.2. Laboratory Experimental Assessments

The results obtained by Boccardi et al. [75] suggested that the throttling losses of a R744 air-to-water heat pump unit can be decreased by 46% by adopting the multi-ejector concept over the investigated running modes. However, the improvement in overall exergy efficiency was found to be, at best, equal to 9% in comparison with the basic solution.

Boccardi et al. [76] performed a sensitivity analysis on a multi-ejector CO_2 heat pump water heater (see Figure 28) considering the ejector area ratio, the compressor frequency and the outdoor temperature. The outcomes obtained showed the existence of an optimal multi-ejector configuration, as depicted in Figure 29a. In addition, the evaluation revealed that the optimum ejector performance does not coincide with the best performance in terms of *COP* and heating capacity (Figure 29b). This implies that the performance of the ejector can be enhanced by improving its design. Finally, it was found that it is needed to switch from an ejector configuration to another in order to maximize the performance with respect to the external temperature (Figure 29c).

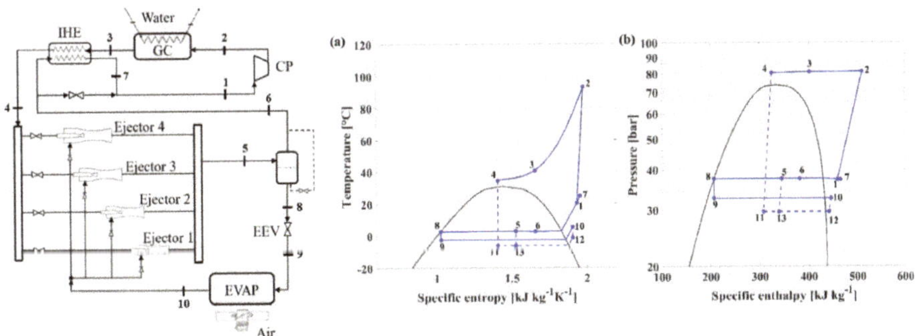

Figure 28. Schematic of the multi-ejector R744 air-to-water heat pump unit (left-hand side) and its T-s (**a**) and p-h (**b**) diagrams [76].

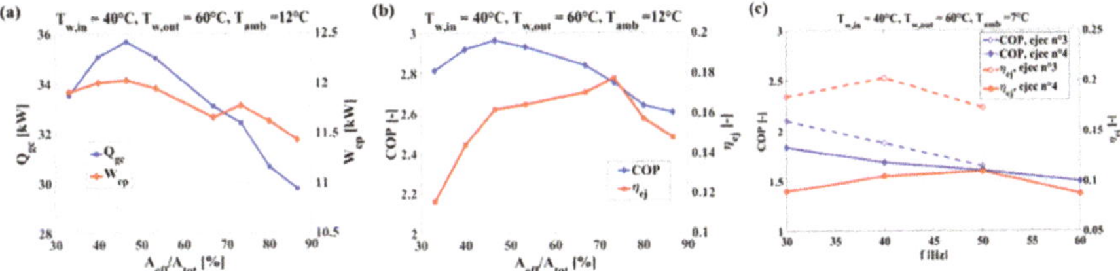

Figure 29. (**a**) Heating capacity (left) and compressor work (right) variations as a function of the overall ejector cross section; (**b**) *COP* (left) and ejector efficiency (right) variations as a function of the overall ejector cross section; (**c**) *COP* (left) and ejector efficiency (right) as a function of the compressor frequency [76].

Boccardi et al. [75,76] also evaluated ejector efficiencies below the ones available in the open literature. This was due to the fact that a multi-ejector module designed for refrigeration applications (i.e., high pressure lift and low mass entrainment ratio) was used.

Singh et al. [77] experimentally assessed the performance of a transcritical R744 heat pumping system equipped with multi-ejector block aimed at hot water production in the Indian climate context.

The results showed that total *COP*s between 7.2 and about 4 and *COP*s in cooling mode between 4 and about 2 can be achieved at R744 gas cooler outlet temperatures from 36 °C and 46 °C. The evaporating temperature was ranged between 7 °C and 9 °C during the experimental campaign.

5. Conclusions and Future Work

The development of the multi-ejector concept has helped to consolidate the position of commercial transcritical CO_2 refrigeration systems as market-ready alternatives to HFCs in any climate context. Also, their reliability and feasibility have been widely proved via a relevant number of installations and considerable energy savings have been estimated as well as actual energy consumptions have been monitored and reported in the literature. However, many retailers are still reluctant to consider R744 as the sole refrigerant for supermarkets located in warm climates due to:

(1) The persevering non-technological barriers, amongst which the lack of awareness of available technologies at decision making level and the lack of trained installers and service technicians [78];
(2) The limited amount of available field measurements and economic evaluations, especially with respect to the latest proposed solutions (i.e., Units relying on two multi-ejector blocks and/or implementing direct heating and cooling fan coils and air curtains). The availability of such information would help to build confidence in these solutions and thus lead to finally open the doors to their market penetration in warm climates as well.

Also, the multi-ejector concept is perceived to be the key to promote "CO_2 only" vapor-compression systems in other high energy-demanding buildings (e.g., hotels, spas, gyms) located in warm locations. However, the implementation of multiple ejectors operating in parallel is at an early development stage in this sector. The proliferation of these solutions will heavily depend on the energy advantageous attainable by simultaneously adopting a multi-ejector block and overcoming the secondary fluid penalization via direct heating and cooling technique, especially in warm areas.

The aforementioned knowledge gap will be bridged with the aid of MultiPACK (www.ntnu.edu/multipack).

Finally, it is worth remarking that the multi-ejector concept is still in its infancy in high ambient temperature countries (e.g., India) and extensive energy, environmental and economic assessments based on field measurements are necessary.

Author Contributions: Conceptualization, P.G.; Investigation, P.G.; Resources, P.G.; Writing—Original Draft Preparation, P.G.; Writing—Review & Editing, A.H., K.B., S.M. and E.E.K.; Supervision, A.H., K.B., S.M. and E.E.K.

Funding: The research leading to these results has received funding from the European Union under the programme H2020-EU.3.3.1.—Reducing energy consumption and carbon footprint by smart and sustainable use (Grant Agreement number: 723137).

Conflicts of Interest: The authors declare no conflict of interest.

Nomenclature

Symbols, abbreviations and subscripts/superscripts

AC	Air conditioning
COP	Coefficient of Performance (-)
DHW	Domestic hot water
GWP	Global Warming Potential ($kg_{CO_{2,equivalent}} \cdot kg^{-1}_{refrigerant}$)
h	Enthalpy per unit of mass ($kJ \cdot kg^{-1}$)
HCFC	Hydrochlorofluorocarbon
HFC	Hydrofluorocarbon
HP	High pressure (bar)
HVP	High pressure electronic expansion valve

Symbols, abbreviations and subscripts/superscripts

HX	Heat exchanger
IHX	Internal heat exchanger
IP	Intermediate pressure (bar)
LP	Low pressure (bar)
LT	Low temperature (°C)
$\dot{m}$	Mass flow rate (kg·s^{-1})
MP	Medium pressure (bar)
MT	Medium temperature (°C)
ODP	Ozone Depletion Potential
p	Pressure (bar)
$\dot{Q}$	Heat transfer rate (kW)
RC&H	Refrigeration, cooling & heating
s	Entropy per unit of mass (kJ·kg^{-1}·K^{-1})
t	Temperature (°C)
TEWI	Total Equivalent Warming Impact (ton$_{CO2,equivalent}$)
$\dot{W}$	Power (kW)

Greek symbols

η	Efficiency [-]
Π	Suction pressure ratio [-]
ω	Mass entrainment ratio [-]

References

1. European Commission. *Regulation (EU) No 517/2014 of the European Parliament and of the Council of 16th April 2014 on Fluorinated Greenhouse Gases and Repealing Regulation (EC) No 842/2006*; EU: Brussels, Belgium, 2014.
2. International Institute of Refrigeration. *Newsletter No. 73*; International Institute of Refrigeration: Paris, France, 2018.
3. UNEP. *Report of the Twenty-Eighth Meeting of the Parties to the Montreal Protocol on Substances that Deplete the Ozone Layer*; UNEP: Kigali, Rwanda, 2016.
4. Ciconkov, R. Refrigerants: There is still no vision for sustainable solutions. *Int. J. Refrig.* **2018**, *86*, 441–448. [CrossRef]
5. Ge, Y.T.; Tassou, S.A. Thermodynamic analysis of transcritical CO_2 booster refrigeration systems in supermarket. *Energy Convers. Manag.* **2011**, *52*, 1868–1875. [CrossRef]
6. Hafner, A. 2020 perspectives CO_2 refrigeration and heat pump systems. In Proceedings of the 6th IIR Ammonia and CO_2 Refrigeration Technologies Conference, Ohrid, Macedonia, 16–18 April 2015.
7. De Oña, A.; Gkizelis, A.; Skačanová, K.; Boccabella, E. Global market and policy trends for CO_2 and ammonia as natural refrigerants. In Proceedings of the 7th IIR Ammonia and CO_2 Refrigeration Technologies Conference, Ohrid, Macedonia, 11–13 May 2017.
8. Gullo, P.; Hafner, A.; Banasiak, K. Transcritical R744 refrigeration systems for supermarket applications: Current status and future perspectives. *Int. J. Refrig.* **2018**, *93*, 269–310. [CrossRef]
9. Hafner, A.; Poppi, S.; Nekså, P.; Minetto, S.; Eikevik, T.M. Development of commercial refrigeration systems with heat recovery for supermarket building. In Proceedings of the 10th IIR Gustav Lorentzen Conference on Natural Refrigerants, Delft, The Netherlands, 25–27 June 2012.
10. Hafner, A.; Försterling, S.; Banasiak, K. Multi-ejector concept for R-744 supermarket refrigeration. *Int. J. Refrig.* **2014**, *43*, 1–13. [CrossRef]
11. Cavallini, A.; Zilio, C. Carbon dioxide as a natural refrigerant. *Int. J. Low-Carbon Technol.* **2007**, *2*, 225–249. [CrossRef]
12. Elbel, S.; Lawrence, N. Review of recent developments in advanced ejector technology. *Int. J. Refrig.* **2016**, *62*, 1–18. [CrossRef]
13. Banasiak, K.; Hafner, A.; Haddal, O.; Eikevik, T. Test facility for a multiejector R744 refrigeration system. In Proceedings of the 11th IIR Gustav Lorentzen Conference on Natural Refrigerants, Hangzhou, China, 31 August–2 September 2014.

14. Elbel, S.; Hrnjak, P. Experimental validation of a prototype ejector design to reduce throttling losses encountered in transcritical R744 system operate. *Int. J. Refrig.* **2008**, *31*, 411–422. [CrossRef]

15. Banasiak, K.; Hafner, A.; Kriezi, E.E.; Madsen, K.B.; Birkelund, M.; Fredslund, K.; Olsson, R. Development and performance mapping of a multi-ejector expansion work recovery pack for R744 vapour compression units. *Int. J. Refrig.* **2015**, *57*, 265–276. [CrossRef]

16. Sever, M. Transcritical CO_2 refrigeration systems in all climates. In Proceedings of the ATMOsphere Japan 2018, Tokyo, Japan, 13 February 2018.

17. Girotto, S. Improved transcritical CO_2 refrigeration systems for warm climates. In Proceedings of the 7th IIR Ammonia and CO_2 Refrigeration Technologies Conference, Ohrid, Macedonia, 11–13 May 2017.

18. Marimon, M.A.; Aria, J.; Lundqvist, P.; Bruno, J.C.; Coronas, A. Integration of trigeneration in an indirect cascade refrigeration system in supermarkets. *Energy Build.* **2011**, *43*, 1427–1434. [CrossRef]

19. Karampour, M.; Sawalha, S.; Arias, J. Eco-Friendly Supermket—An Overview, Public Report 2 for SuperSmart Project (Project Number: 696076). 2016. Available online: http://www.supersmart-supermarket.info/downloads/ (accessed on 12 December 2018).

20. Nekså, P.; Hafner, A.; Bredesen, A.; Eikevik, T.M. CO_2 as working fluid—Technological development on the road to sustainable refrigeration. In Proceedings of the 12th IIR Gustav Lorentzen Natural Working Fluids Conference, Edinburgh, UK, 21–24 August 2016.

21. Hafner, A.; Banasiak, K.; Fredslund, K.; Girotto, S.; Smolka, J. R744 ejector system case: Italian supermarket, Spiazzo. In Proceedings of the 12th IIR Gustav Lorentzen Natural Working Fluids Conference, Edinburgh, UK, 21–24 August 2016.

22. Sawalha, S. Investigation of heat recovery in CO_2 trans-critical solution for supermarket refrigeration. *Int. J. Refrig.* **2013**, *36*, 145–156. [CrossRef]

23. Ge, Y.T.; Tassou, S.A. Performance evaluation and optimal design of supermarket refrigeration systems with supermarket model "SuperSim". Part II: Model applications. *Int. J. Refrig.* **2011**, *34*, 540–549. [CrossRef]

24. Ommen, T.; Elmegaard, B. Numerical model for thermoeconomic diagnosis in commercial transcritical/subcritical booster refrigeration systems. *Energy Convers. Manag.* **2012**, *60*, 161–169. [CrossRef]

25. Sawalha, S.; Piscopiello, S.; Karampour, M.; Tamilarasan, M.L.; Rogstam, J. Field Measurements of Supermarket Refrigeration Systems. Part II: Analysis of HFC refrigeration systems and comparison to CO_2 trans-critical. *Appl. Therm. Eng.* **2017**, *111*, 170–182. [CrossRef]

26. Karampour, M.; Sawalha, S. Energy efficiency evaluation of integrated CO_2 trans-critical system in supermarkets: A field measurements and modelling analysis. *Int. J. Refrig* **2017**, *82*, 470–486. [CrossRef]

27. Purohit, N.; Gullo, P.; Dasgupta, M.S. Comparative assessment of low-GWP based refrigerating plants operating in hot climates. *Energy Procedia* **2017**, *109*, 138–145. [CrossRef]

28. Gullo, P.; Elmegaard, B.; Cortella, G. Advanced exergy analysis of a R744 booster refrigeration system with parallel compression. *Energy* **2016**, *107*, 562–571. [CrossRef]

29. Gullo, P.; Elmegaard, B.; Cortella, G. Energy and environmental performance assessment of R744 booster supermarket refrigeration systems operating in warm climates. *Int. J. Refrig.* **2016**, *64*, 61–79. [CrossRef]

30. Minetto, S.; Brignoli, R.; Zilio, C.; Marinetti, S. Experimental analysis of a new method of overfeeding multiple evaporators in refrigeration systems. *Int. J. Refrig.* **2014**, *38*, 1–9. [CrossRef]

31. Gullo, P.; Cortella, G.; Minetto, S.; Polzot, A. Overfed evaporators and parallel compression in commercial R744 booster refrigeration systems—An assessment of energy benefits. In Proceedings of the 12th IIR Gustav Lorentzen Natural Working Fluids Conference, Edinburgh, UK, 21–24 August 2016.

32. Polzot, A.; D'Agaro, P.; Gullo, P.; Cortella, G. Modelling commercial refrigeration systems coupled with water storage to improve energy efficiency and perform heat recovery. *Int. J. Refrig.* **2016**, *69*, 313–323. [CrossRef]

33. Polzot, A.; Gullo, P.; D'Agaro, P.; Cortella, G. Performance evaluation of a R744 booster system for supermarket refrigeration, heating and DHW. In Proceedings of the 12th IIR Gustav Lorentzen Natural Working Fluids Conference, Edinburgh, UK, 21–24 August 2016.

34. Ge, Y.T.; Tassou, S.A. Control optimizations for heat recovery from CO_2 refrigeration systems in supermarket. *Energy Convers. Manag.* **2014**, *78*, 245–252. [CrossRef]

35. Reinholdt, L.; Madsen, C. Heat recovery on CO_2 systems in supermarkets. In Proceedings of the 9th IIR Gustav Lorentzen Conference on Natural Refrigerants, Sydney, Australia, 12–14 April 2010.

36. Minetto, S.; Girotto, S.; Rossetti, A.; Marinetti, S. Experience with ejector work recovery and auxiliary compressors in CO_2 refrigeration systems. Technological aspects and application perspectives. In Proceedings of the 6th IIR Ammonia and CO_2 Refrigeration Technologies Conference, Ohrid, Macedonia, 16–18 April 2015.

37. Schönenberger, J. Experience with R744 refrigerating systems and implemented multi ejectors and liquid overfeed. In Proceedings of the 12th IIR Gustav Lorentzen Natural Working Fluids Conference, Edinburgh, UK, 21–24 August 2016.

38. Hafner, A.; Banasiak, K. Full scale supermarket laboratory R744 ejector supported and AC integrated parallel compression unit. In Proceedings of the 12th IIR Gustav Lorentzen Natural Working Fluids Conference, Edinburgh, UK, 21–24 August 2016.

39. Hafner, A.; Fredslund, K.; Banasiak, K. Next generation R744 refrigeration technology for supermarkets. In Proceedings of the 24th IIR International Congress of Refrigeration, Yokohama, Japan, 16–22 August 2015.

40. Hafner, A. Integrated CO_2 system refrigeration, air conditioning and sanitary hot water. In Proceedings of the 7th IIR Ammonia and CO_2 Refrigeration Technologies Conference, Ohrid, Macedonia, 11–13 May 2017.

41. Girotto, S. Direct space heating and cooling with CO_2 refrigerant—A new solution for commercial buildings. In Proceedings of the ATMOsphere Europe 2016, Barcelona, Spain, 19–20 April 2016.

42. Schönenberger, J.; Hafner, A.; Banasiak, K.; Girotto, S. Experience with ejectors implemented in a R744 booster system operating in a supermarket. In Proceedings of the 11th IIR Gustav Lorentzen Conference on Natural Refrigerants, Hangzhou, China, 31 August–2 September 2014.

43. Hafner, A.; Schönenberger, J.; Banasiak, K.; Girotto, S. R744 ejector supported parallel vapour compression system. In Proceedings of the 3rd IIR International Conference on Sustainability and Cold Chain, London, UK, 23–25 June 2014.

44. Minetto, S.; Girotto, S.; Salvatore, M.; Rossetti, A.; Marinetti, S. Recent installations of CO_2 supermarket refrigeration system for warm climates: Data from field. In Proceedings of the 3rd IIR International Conference on Sustainability and Cold Chain, London, UK, 23–25 June 2014.

45. Javerschek, O.; Reichle, M.; Karbiner, J. Influence of ejectors on the selection of compressors in carbon dioxide booster systems. In Proceedings of the 9th International Conference on Compressors and Coolants, Bratislava, Slovakia, 6–8 September 2017.

46. Bodys, J.; Palacz, M.; Haida, M.; Smolka, J.; Nowak, A.J.; Banasiak, K.; Hafner, A. Full-scale multi-ejector module for a carbon dioxide supermarket refrigeration system. Numerical study of performance evaluation. *Energy Convers. Manag.* **2017**, *138*, 312–326. [CrossRef]

47. Pisano, G. The use of ejectors technology: How to boost efficiency in warm climates—A real example from Italy. In Proceedings of the 13th IIR Gustav Lorentzen Conference on Natural Refrigerants, Valencia, Spain, 18–20 June 2018.

48. Gullo, P.; Hafner, A.; Cortella, G. Multi-ejector R744 booster refrigerating plant and air conditioning system integration—A theoretical evaluation of energy benefits for supermarket applications. *Int. J. Refrig.* **2017**, *75*, 164–176. [CrossRef]

49. Gullo, P.; Tsamos, K.; Hafner, A.; Ge, Y.; Tassou, S. State-of-the-art technologies for R744 refrigeration systems—A theoretical assessment of energy advantages for European food retail industry. *Energy Procedia* **2017**, *123*, 46–53. [CrossRef]

50. European Commission. *ANNEXES to the REPORT FROM THE COMMISSION Assessing the 2022 Requirement to Avoid Highly Global Warming Hydrofluorocarbons in Some Commercial Refrigeration Systems*; C(2017) 5230 Final ANNEXES 1 to 2; European Commission: Brussels, Belgium, 2017.

51. Gullo, P.; Hafner, A. Comparative assessment of supermarket refrigeration systems using ultra low-GWP refrigerants—Case study of selected American cities. In Proceedings of the 30th International Conference on Efficiency, Cost, Optimisation, Simulation and Environmental Impact of Energy Systems, San Diego, CA, USA, 2–6 July 2017.

52. Gullo, P.; Tsamos, K.M.; Hafner, A.; Banasiak, K.; Ge, Y.T.; Tassou, S.A. Crossing CO_2 equator with the aid of multi-ejector concept: A comprehensive energy and environmental comparative study. *Energy* **2018**, *164*, 236–263. [CrossRef]

53. Madsen, K.B.; Kriezi, E.K. Financial aspects of ejector solutions in supermarket and smaller industrial systems. In Proceedings of the 13th IIR Gustav Lorentzen Conference on Natural Refrigerants, Valencia, Spain, 18–20 June 2018.

54. Gullo, P.; Hafner, A.; Banasiak, K. Thermodynamic Performance Investigation of Commercial R744 Booster Refrigeration Plants Based on Advanced Exergy Analysis. *Energies* **2019**, *12*, 354. [CrossRef]

55. Fredslund, K.; Kriezi, E.E.; Madsen, K.B.; Birkelund, M.; Olsson, R. CO_2 installations with a multi ejector for supermarkets, case studies from various locations. In Proceedings of the 12th IIR Gustav Lorentzen Natural Working Fluids Conference, Edinburgh, UK, 21–24 August 2016.

56. Haida, M.; Banasiak, K.; Smolka, J.; Hafner, A.; Eikevik, T.M. Experimental analysis of the R744 vapour compression rack equipped with the multi-ejector expansion work recovery module. *Int. J. Refrig.* **2016**, *64*, 93–107. [CrossRef]

57. Pardiñas, A.A.; Hafner, A.; Banasiak, K. Integrated R744 ejector supported parallel compression racks for supermarkets. Experimental results. In Proceedings of the 13th IIR Gustav Lorentzen Conference on Natural Refrigerants, Valencia, Spain, 18–20 June 2018.

58. Hafner, A.; Hemmingsen, A.K. R744 refrigeration technologies for supermarkets in warm climates. In Proceedings of the 24th IIR International Congress of Refrigeration, Yokohama, Japan, 16–22 August 2015.

59. Kriezi, E.E.; Fredslund, K.; Birkelund, M.; Banasiak, K.; Hafner, A. R744 multi ejector development. In Proceedings of the 12th IIR Gustav Lorentzen Natural Working Fluids Conference, Edinburgh, UK, 21–24 August 2016.

60. Kriezi, E.E.; Larsen, L.F.S.; Piscopiello, S.; Madsen, K.B. System efficiency and energy savings with parallel compression and ejectors under flooded conditions. In Proceedings of the 13th IIR Gustav Lorentzen Conference on Natural Refrigerants, Valencia, Spain, 18–20 June 2018.

61. Pardiñas, A.A.; Hafner, A.; Banasiak, K.; Kvalsik, K.H.; Larsen, L. Strategy for the control of two groups of ejectors operating in parallel in integrated R744 refrigeration systems. In Proceedings of the 13th IIR Gustav Lorentzen Conference on Natural Refrigerants, Valencia, Spain, 18–20 June 2018.

62. Pardiñas, A.A.; Hafner, A.; Banasiak, K. Novel integrated CO_2 vapour compression racks for supermarkets. Thermodynamic analysis of possible system configurations and influence of operational conditions. *Appl. Therm. Eng.* **2018**, *131*, 1008–1025. [CrossRef]

63. Isaac, M.; van Vuuren, D.P. Modeling global residential sector energy demand for heating and air conditioning in the context of climate change. *Energy Policy* **2009**, *37*, 507–521. [CrossRef]

64. Middle East's First CO_2 Supermarket Opens in Jordan. Available online: http://r744.com/articles/8148/middle_east_first_co2_supermarket_opens_in_jordan?utm_source=mailchimp&utm_medium=email&utm_campaign=Bi-weekly%20Newsletter (accessed on 12 December 2018).

65. Kvalsvik, K.H.; Banasiak, K.; Hafner, A. Integrated CO_2 refrigeration and AC unit for hot climates. In Proceedings of the 7th IIR Ammonia and CO_2 Refrigeration Technologies Conference, Ohrid, Macedonia, 11–13 May 2017.

66. Singh, S.; Amshith, R.; Prakash, M.M.; Banasiak, K.; Hafner, A.; Nekså, P. Analysis of R744 refrigeration system with liquid ejectors. In Proceedings of the 13th IIR Gustav Lorentzen Conference on Natural Refrigerants, Valencia, Spain, 18–20 June 2018.

67. Singh, S.; Banasiak, K.; Hafner, A.; Maiya, P.; Nekså, P. Performance evaluation of CO_2 ejector system with parallel compressor for supermarket application. In Proceedings of the 17th International Refrigeration and Air Conditioning Conference at Purdue, West Lafayette, IN, USA, 9–12 July 2018.

68. Blust, S.; Singh, S.; Hafner, A.; Banasiak, K.; Nekså, P. Environment-friendly refrigeration packs for Indian supermarkets: Experimental investigation of energy performance of a multiejector-driven R744 integrated compressor rack. In Proceedings of the 13th IIR Gustav Lorentzen Conference on Natural Refrigerants, Valencia, Spain, 18–20 June 2018.

69. Singh, S.; Hafner, A.; Banasiak, K.; Maiya, P.; Nekså, P. Experimental evaluation of multi-ejector based CO_2 cooling system for supermarkets in tropical zones. In Proceedings of the 17th International Refrigeration and Air Conditioning Conference at Purdue, West Lafayette, IN, USA, 9–12 July 2018.

70. Nekså, P.; Rekstad, H.; Zakeri, G.; Schiefloe, P.A. CO_2-heat pump water heater: Characteristics, system design and experimental results. *Int. J. Refrig.* **1998**, *21*, 172–179. [CrossRef]

71. Nekså, P. CO_2 heat pump systems. *Int. J. Refrig.* **2002**, *25*, 421–427. [CrossRef]

72. Cecchinato, L.; Chiarello, M.; Corradi, M. Design and experimental analysis of a carbon dioxide transcritical chiller for commercial refrigeration. *Appl. Energy* **2010**, *87*, 2095–2101. [CrossRef]

73. Minetto, S.; Cecchinato, L.; Brignoli, R.; Marinetti, S.; Rossetti, A. Water-side reversible CO_2 heat pump for residential application. *Int. J. Refrig.* **2016**, *63*, 237–250. [CrossRef]

74. Schoenenberger, J.; Fraga, G.R. Experience with industrial refrigerating systems for southern climate. In Proceedings of the 13th IIR Gustav Lorentzen Conference on Natural Refrigerants, Valencia, Spain, 18–20 June 2018.
75. Boccardi, G.; Botticella, F.; Lillo, G.; Mastrullo, R.; Mauro, A.W.; Trinchieri, R. Thermodynamic Analysis of a Multi-Ejector, CO_2, Air-To-Water Heat Pump System. *Energy Procedia* **2016**, *101*, 846–853. [CrossRef]
76. Boccardi, G.; Botticella, F.; Lillo, G.; Mastrullo, R.; Mauro, A.W.; Trinchieri, R. Experimental investigation on the performance of a transcritical CO_2 heat pump with multi-ejector expansion system. *Int. J. Refrig.* **2017**, *82*, 389–400. [CrossRef]
77. Singh, S.; Amshith, R.; Prakash, M.M.; Banasiak, K.; Hafner, A.; Nekså, P. Performance Investigation of a Multi-ejector R744 Heat Pump. In Proceedings of the 13th IIR Gustav Lorentzen Conference on Natural Refrigerants, Valencia, Spain, 18–20 June 2018.
78. Minetto, S.; Marinetti, S.; Saglia, P.; Masson, N.; Rossetti, A. Non-technological barriers to the diffusion of energy-efficient HVAC&R solutions in the food retail sector. *Int. J. Refrig.* **2018**, *86*, 422–434. [CrossRef]

Review

Control Strategies in Multi-Zone Air Conditioning Systems

Behzad Rismanchi [1,*], Juan Mahecha Zambrano [1], Bryan Saxby [2], Ross Tuck [2] and Mark Stenning [2]

1   Department of Infrastructure Engineering, The University of Melbourne, Melbourne 3010, Australia; jmahecha@student.unimelb.edu.au
2   OAIRO Alliance Ltd, Unit M, Bourne End Business Park, Bucks SL66TG, UK; bryan@oairo.com (B.S.); ross@oairo.com (R.T.); mark@oairo.com (M.S.)
*   Correspondence: brismanchi@unimelb.edu.au; Tel.: +61-3-8344-1127

Received: 11 December 2018; Accepted: 18 January 2019; Published: 23 January 2019

Abstract: In a commercial building, a significant amount of energy is used by the ventilation systems to condition the air for the indoor environments to satisfy the required quantity (temperature and humidity) and quality (amount of fresh air). For many years, Variable Air Volume (VAV) systems have been considered as the most efficient solutions by balancing the airflow volume based on the demand making them energy efficient when compared with the traditional Constant Air Volume (CAV) systems. However, the setpoints in VAV systems are often misread by the sensors due to stratification and formation of pollutant pockets and responding to design levels that overestimate the real-time demand conditions, which result in waste of energy, thermal discomfort and unhealthy air. In general, VAV devices are expensive, complicated and prone to failures and they are used only in medium and large projects. More recently, new technologies have evolved to solve this issue. In one of the new solutions, VAV motors terminals are replaced with flaps which are simpler and less expensive thus, they can be implemented in a wider range of projects. In systems, balancing and supplying the optimal airflow to reduce the energy consumption while delivering ideal thermal and Indoor Air Quality (IAQ) levels are the main challenges. In this paper, a comparison of the recent technologies with traditional VAV systems is presented to be used as a guild line for researchers and designers in the field of Heating Ventilation Air Conditioning (HVAC).

Keywords: HVAC; pressure based control; damper control; static pressure reset; CO_2 reset; demand-based control; energy saving; human well-being; IAQ; Atomic Air

---

1. Introduction

Climate change, air pollution and global warming are the critical concerns that need immediate action. Even though mitigation measures have been implemented around the world, significant efforts are still required to limit the rise in the global temperature to the 2 °C as stated in the Paris Agreement [1–3]. Building sector accounts for approximately 40% of the total world final energy consumption and around one-third of the greenhouse emissions, therefore, it plays an important role to reduce the impact on the environment [4–6].

In buildings, the major portion of energy is consumed by the mechanical services, heating and air conditioning systems, with the share of around 50–82% [7,8]. Consequently, during the last decade researchers have focused on the optimisation and development of strategies to improve the efficiency of such systems [9,10]. However, reducing the amount of energy consumed by the air conditioning systems, must not compromise the indoor air quality (IAQ) and thermal comfort [9]. Because as it has been shown, IAQ has a direct impact not only on people's health and wellbeing but also on their productivity [11] as they spend approximately 90% of their time within indoor environments [12,13].

Typically, ventilation in buildings is provided using heating, ventilation and air conditioning systems (HVAC) that condition air by utilizing either indoor conditioning units that recycle indoor air or centralised air handling units (AHU) that may utilize either indoor or outdoor air and then supply the conditioned air into each zone. Since multiple zones could have different requirements due to differences in the number of occupants, equipment, devices and activities, the ventilation in different terminals may be unevenly distributed producing discomfort. Also, energy could be wasted in over ventilated spaces due to poor equipment control, variability, supply and exhaust fans typically work on an equal basis and work more than required [14].

To address these challenges and reduce the energy consumption by the ventilation system, it is identified that the air distribution system is the critical parameter. Reducing the volume of airflow required by the system, while achieving the appropriate levels of IAQ, will have a significant impact on the overall performance of the HVAC systems [15]. Equally important, if the ventilation system is using less air flow—so that conditioning less air—equipment such as the chiller, cooling towers, pumps and other mechanical services will have a lower load with less energy demand.

Early HVAC systems were based on constant air volume (CAV) controls that condition the spaces by heating/cooling the constant airflow depending on the requirement thus they can only control the temperature of a specific area. Consequently, CAV systems cannot satisfy the temperature requirements in every zone. The variable air volume system (VAV) was later introduced as a novel solution to control the air flow rate in responding to dynamic loads in each zone [16]. In the terminals of the system, VAV boxes measure and control the volume air flow by opening or closing dampers to deliver the required amount of air. Then, the change in the pressure drop in the air ducts, generated by the dampers (30%–100%), triggers a signal to the supply air fan to increase or decrease the air flow [10]. However, despite that the VAV systems are efficient, their performance highly depends on their control strategy [17].

The VAV systems have been implemented since the end of the World War II and became highly popular in the 70's. Since then, various improvements in the control, measurement techniques and sensors have been developed [18]. More recently different studies have been looking for simpler, more efficient and less expensive solutions to deliver the minimum amount of airflow while endeavouring to provide an ideal IAQ [15]. By using motor flaps—instead of VAV boxes—the air flow could be regulated. However, determining the current airflow becomes a challenge since it cannot be measured directly without significant cost implications. Therefore, the air flow as a direct control parameter must be replaced, for example, by the zone pressure, the static duct pressure, the temperature, CO_2 level or a combination of them.

Until now, no attempt has been made to compare the new control strategies for minimising the air flow in multi-zone air conditioning systems. This paper mainly focusses on the investigation and comparison of these control strategies, their control objectives and their control parameters. The main goal is to identify the energy saving potential of each alternative and their impact on the indoor environment. Consequently, the structure of the paper is as follows: Section 2 describes in more detail the problem and the motivation of this paper. Section 3 focused on developments related to VAV systems and their control strategies; Section 4 discussed and compared the modern control strategies to assess their impact on the thermal comfort, IAQ and energy consumption.

2. Background

The focus of this literature survey is on modern control strategies that relay on enhancing the balance of airflow within multiple zones to improve the energy consumption of the HVAC system. Furthermore, it focuses on the assessment of each control strategy looking at their impact on the thermal comfort as well as IAQ. Figure 1 shows a schematic diagram of a typical VAV system. In general, these systems have a central 1–3 speed AHU that delivers primary air at a defined temperature to terminal boxes in each zone. The terminal boxes or VAV boxes have a primary-air damper controlled automatically. This damper regulates the volume of primary air delivered to the box according to the demand. Since each box regulates the air flow independently, the total volume supplied by the AHU varies according to the demand of all the boxes. Thus, the variable airflow is achieved by controlling the speed of the supply air fan [19].

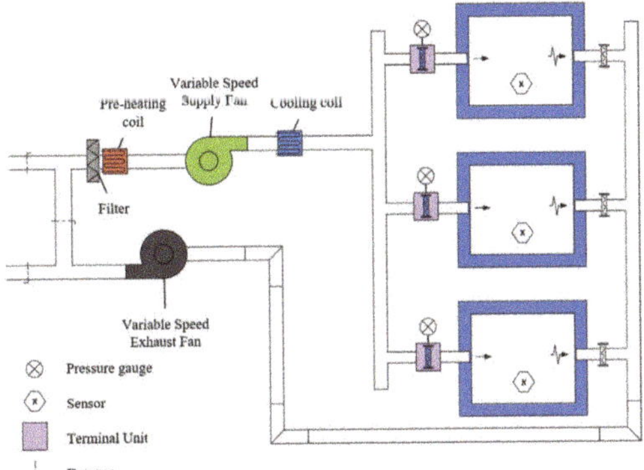

Figure 1. Schematic diagram of a typical variable air volume (VAV) system. Adapted from [20].

Regarding the energy consumption of the ventilation system, between all the components, the supply and exhaust fans are responsible for 50% and 33%, respectively [21] hence, conservation measures should be focused in these devices. In addition, it can be demonstrated that the air volume flow has quadratic and cubic relations with the pressure and the power consumption correspondingly. This means that a reduction of the air volume flow by 10% will result in power consumption savings of 27%. In detail, it is critical to reduce the overall air volume flow required by the VAV system to meet the comfort levels as well as avoiding peak periods of high air volume demand followed by low air volume flow demand periods [15].

In conventional systems, VAV boxes play a major role in both energy consumption and thermal comfort as they control the overall airflow required (but struggle with and are restricted to a 30% minimum supply to avoid airdrop) by the system as well as the airflow delivered to each zone. They have multiple components such as a controller, temperature sensor, actuator, damper, reheating coil and air flow sensor [22], VAV boxes are complex devices with embedded electronics which are expensive and prone to failures [15,19]. Hence, they are only applicable to large or medium construction projects [7].

Even though VAV systems are efficient solutions, it is still difficult to minimise the air volume set point. The simplest control solution is to set a constant air flow or a constant static pressure in the air ducts—usually during commissioning—for one specific occupancy scheme. The VAV terminals adjust the damper to deliver the required amount of air in each zone. Then, the supply fan adjusts the air flow to maintain the setpoint. However, often the set points are higher than required because they are not adjustable to the real demand, causing discomfort and waste of energy. As a consequence, many researchers have been focusing their efforts in developing complex control strategies such as Static Pressure Reset (SPR) [18,23], CO_2 reset [7,24,25], time average controls [26], fault adaptive controls [19], etc. These strategies are based on empirical, physics-based, data-driven or grey box approaches which are complex and sometimes difficult to implement in existing systems. Additionally, novel strategies are exploring a physical phenomenon known as diffuse ventilation to improve the environmental indoor climate and, at the same time, reduce the energy consumption.

Furthermore, with the aim of reducing complexity, extending the usage of VAV systems to small and medium projects, reducing the costs and ensuring a more efficient usage of the energy as well as a proper IAQ and thermal comfort, different studies have focused in the development or adaptation of control strategies to use motor flaps instead of VAV boxes [15,27]. The primary challenge here is how to estimate and control the air flow without measuring it directly—as the VAV boxes attempt to do.

3. Control Strategies in VAV Systems

VAV systems are more efficient than the constant air volume (CAV) systems because they adjust the amount of air supplied in different zones (with different requirements). However, their efficiency highly depends on the control strategy implemented to balance and distribute the airflow in each conditioned space. Early versions of manufacturers' control strategies require to define the principal air flow supply volume as a design parameter (usually defined during the commissioning stage of the project) without being determined automatically during the operation. Commonly, this parameter is overestimated so that the temperature in the zones that are being conditioned are normally lower than required (causing discomfort) and the fans work more than necessary (causing energy wastage). This category of control strategies are summarised by Pang et al. [17] as follows:

- Occupied zone set-point temperatures and night set back
- VAV box minimum flow (typically 30%)
- Optimum start
- Supply air temperature reset
- Economiser and minimum outdoor air intake

These strategies could be improved since in general they are based on maintaining a constant static pressure (CSP) set-point in the main duct without considering the actual pressure demand. In the following sections different strategies are presented. Figure 2 demonstrate the alternatives and features of different VAV systems.

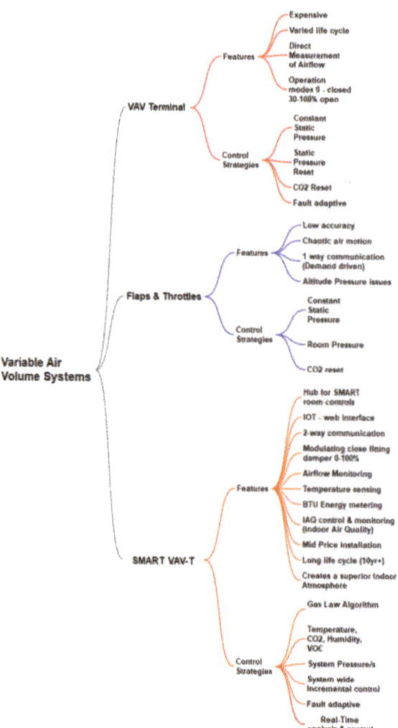

Figure 2. VAV alternatives and features.

3.1. Duct Static Pressure-Based Control Strategies

To reduce the supply fan's energy consumption, different studies have focused on strategies to reset the static pressure setpoint. Early versions of this strategy are based on measuring the airflow rate and mathematical models to determine the static pressure reset setpoint [16,28]. The effectiveness and performance of these solutions are limited by the accuracy of the airflow sensor thus the airflow as a direct feedback parameter is replaced by the position of the VAV boxes damper. This improves the accuracy of the control but increases its complexity as the required mathematical model to describe the relationship between the duct static pressure and the damper position needs to address non-linear and multivariate phenomena with long delay times [29]. For this reason, a solution focused on developing fuzzy control strategies that replace the mathematical models have emerged. The first alternative that is recognised to be more efficient and stable in its category is the Trim and Respond static pressure reset (SPR) control strategy. Koulani et al. [30] studied on its operation and potentials by developing a mathematical model using Simulink and validating it with an experimental setup. This method involves the integration of a building management system (BMS) with the VAV terminals by informing the supply fan with the pressure needed to satisfy the conditions in the most critical zones. Particularly, the Trim & Respond SPR method is based on zone pressure request alarms. In this case, every damper of the VAV system sends an alarm signal when it is open more than 85% and keeps sending the signal until the damper closes to 80%. Additionally, this method resets the pressure set point every 90 sec within a specific pressure range—upper limit is equal to the CSP setpoint and the lower limit is determined according to the pressure demand to guarantee an optimal damper operation. Koulani et al. concluded by simulating the SPR and CSP control strategies and calculating the fan power that the energy consumption is reduced by 14%. However, they did not consider the IAQ parameters in the study.

Zhang et al. [29] developed a strategy where the zone temperature sensor follows its setpoint by adjusting the supply airflow which is controlled by the terminal damper position feedback. Additionally, two sequential controllers are implemented to realize a pressure independent control. The first one uses the temperature of the conditioned space and its setpoint as the inputs to define the supply airflow setpoint. Then, the second one takes the calculated airflow setpoint and the airflow measure to define the position of the damper. This solution ensures a supply airflow to meet the demand according to occupancies' thermal demand and the indoor load. In particular, this method decreases the terminal damper resistance loss and the system resistance so that a lower static pressure setpoint is achieved as well as better stability.

Walaszczyk and Cichón [18] analysed the impact of not only the trim and response strategy but the proportional-integral-derivative (PID) control on the energy consumption of the HVAC system [18]. The last one involves a standard PID control to determine the static pressure set point. Explicitly, as the zone's demand decreases the zone's damper begin to close. Then, the duct static pressure setpoint is reduced until the critical zone's damper is almost open. Walaszczyk and Chichón simulate both strategies and compared them to the CSP method. They found that strategies that reset the duct's static pressure setpoint have a better performance in terms of energy consumption. Additionally, a small improvement was found using the trim and response alternative over the PID control. Figure 3 illustrates the static pressure reset control strategy based on VAV damper opening.

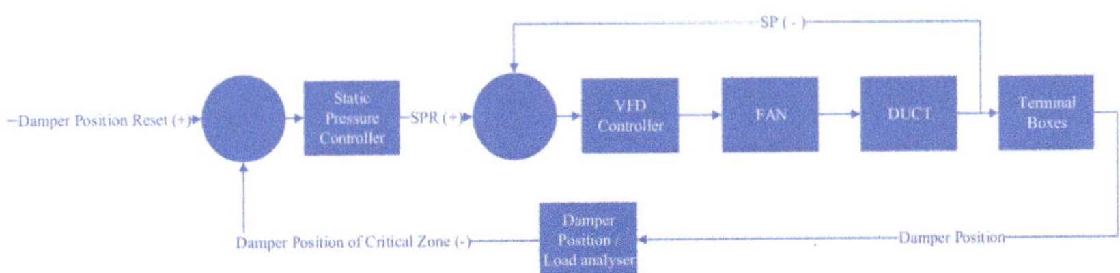

Figure 3. Static pressure reset control strategy based on VAV damper opening [17].

Later, Rahnama et al. [23] developed a SPR strategy that instead of identifying the critical zone by the position of the dampers (i.e., the airflow rate), it measures the static pressure directly at the terminals. This becomes important as they highlighted that not only the air flow is important, but the dimension and layout of the ducting system are determinant factors. They implemented four conditions to control the system and waiting times between actions to let the system reach balance:

- The air flow at the dampers has not changed and the set point is kept—No action required.
- The total air flow increases but the current pressure setpoint is still enough—No action required
- A wide-open damper triggers a request for air from the beginning. The total flow rate is fixed by increasing the pressure set point until the request is fulfilled.
- A partially open damper turns into a wide-open damper and then send a request for air. The total flow rate is fixed by increasing the pressure set point until the request is fulfilled.

Finally, to evaluate this method, Rahnama et al. developed an experimental mock-up laboratory where the air flow rates were assumed to be known at each terminal damper so that there were no real zone areas and demand-controlled ventilation. They found a considerable reduction of 27%, 36% and 21% in fan power use.

Rahnama et al. [31] have continued their investigation on duct static pressure reset control strategies exploring a solution where the terminal dampers are replaced for decentralised fans. In this case, the main supply fan only compensates the pressure of the zone with the lowest pressure drop (critical zone). The decentralised fans provide the required local pressure to fulfil the demand in the other zones. Two sequential controls regulate the speed of the primary and decentralised fans. The last one can be adapted to respond to zone CO_2 levels, occupancy sensors, etc. They implemented this novel solution in an experimental mock-up under different airflow rates. At the end, Rahnama et al. estimated an average reduction of 30% in power use with this new system.

The proliferation of Building Automatization Systems (BMS) made it possible to implement more complex algorithms and exploit a large amount of data that can be recorded. For example, Tukur and Hallinan [32] explored a different approach involving a duct static pressure reset coupled with statistical information. This historical data is used to build a plot to describe the relationship between the system loss coefficient (K) and the weighted mean damper position (D). From the K-D plot, an extensive amount of information to describe the system, can be extracted including: the best damper configurations to deliver a particular condition (quality and quantity of air) and minimise the energy consumption.

More recently, Jing et al. [33] showed that SPR control strategies that use a feedback indicator to reset the static pressure often produce under and over ventilation in different zones. To address this problem, they developed a comprehensive mathematical model to simulate the non-linear phenomena of the ventilation system, this is the relationship between the pressure drop and the airflow at the terminal damper. In addition, they implemented a supervised machine learning algorithm to obtain unknown parameters in the model. Then, they developed a damper position control method to guarantee that the system is well-balanced. Finally, an optimal static-pressure set-point selection method is used to calculate the minimum closed-form static pressure which guarantee the system is energy efficient and delivering the required airflows in each zone. Jing et al validated this strategy and compared it with traditional SPR strategy using an experimental setup. They found energy savings of 21.4% while delivering the desired airflow rates.

3.2. CO_2-Based Control Strategies

Until here, the control strategies presented are based on the duct static pressure to balance the air in different zones taking into account the temperature as an input parameter. Indistinct of the method to balance the air flow, CO_2 could be used to estimate the strength of occupant-related contaminant sources. For example, Lin and Lau [24] proposed a CO_2-based demand control strategy in multi-zone HVAC systems. This is a dynamic reset approach focused on adjusting the outdoor air rate continuously according to the CO_2 produced by the occupants and then modulating the dampers to maintain the outdoor air rate at the new set point. Using CO_2 or occupancy sensors the system reacts to the variation of occupancy rates in one or more zones as well as to the variation of the system ventilation efficiency. Specifically, the CO_2 concentration in each zone, system primary supply and exhaust air, and each zone's primary airflow volume rate must be measured to implement this approach. Then, the outdoor air rate is determined using ANSI/ASHRAE standards 62.1. The performance of this approach is compared with a single path VAV system by using energy and airflow simulations of a classroom/office building modelled with realistic occupancy schedules. It was found that the average annual system outdoor air rate for the proposed approach was 14.6% less than without a demand-controlled ventilation strategy. Additionally, it was estimated that the monetary saving is between 0.3% and 11%.

The previous alternative focused on the system level—since the minimum airflow rate set-point of a VAV boxes is maintained constant based on the design occupancy of the zone. However, a further saving potential can be exploited at the zone level. For this reason, Liu and Lau [25] developed two additional alternatives to dynamically reset the zone airflow to maintain the system either at a target outdoor airflow rate or at a target system ventilation efficiency. On one side, the first option is called CO_2-based DR+ZDR_Vot which has the same previous CO_2 dynamic reset for the system outdoor air rate and it adds, at the zone level, a control that resets the airflow minimum set point to decrease the primary outdoor air fraction by increasing the zone airflow rates in the critical zone. On the other hand, the second option is called CO_2-based DR+ZDR_Ev which reset the minimum set point of the zone primary airflow to maintain the value of the system efficiency greater or equal to the design value. This can be achieved by modulating the zone airflow rate. Again, the system level strategy is the CO_2-based dynamic rest strategy explained previously and the zone level control increases the minimum airflow set-point to decrease the zone outdoor fraction and increase the system ventilation efficiency to a target design efficiency. The same simulations of energy and airflow used in the first part of this study are used to compare and evaluate the performance of these options. The results show that the average annual system outdoor air rates are 44% and 45% less than the values for the VAV system without a demand control strategy. Additionally, the annual monetary savings are between 24% and 46% for the first option and between 26% and 46% for the second option.

Following the CO_2 based controls, Kim et al. [22] proposed a VAV terminal unit control method that meets the ventilation requirement to control IAQ by controlling the air flow rate. The idea is to delay the outdoor air inlet time and to reduce the outdoor air flow rate, to control the CO_2 concentrations in the conditioned spaces. The following actions are proposed if the requirement of CO_2 concentration is not met (less than 1000 ppm):

- Increase the air flow rate of a terminal unit in the critical zone—do not increase the outdoor air flow rate.
- If ventilation CO_2 concentration is low, use recirculated air—do not increase the outdoor air flow rate.
- If above actions are not possible, increase the outdoor air flow rate.

At the end, the minimum airflow rate at the terminal is controlled not only by the temperature of the conditioned zones but their CO_2 concentration to achieve both thermal comfort and IAQ. Additionally, the proposed solution was compared to a fixed minimum air flow method using TRNSYS 17 by reproducing a case where an AHU is selected to condition eight target spaces. It was found that the target temperature is maintained in both cases, but only with the proposed solution, the IAQ was acceptable. Furthermore, the energy consumption was decreased by 20%.

Previous technologies integrate CO_2 based demand controls to guarantee the IAQ together with the thermal comfort and control the air flow using VAV boxes. More recently, different studies are looking for alternatives to the VAV boxes to control the airflow in each zone since they are expensive, complicated, prone to failures and only used in large projects. The challenge of this idea is to find a way to measure and control the air flow without the airflow sensors embedded in the VAV boxes.

Zucker et al. [15] proposed the implementation of motor flaps instead of the VAV boxes. To replace the flow measurement equipment, they developed a model based on the pressure drop of the air ducts and the CO_2 levels to control the flaps positions. In this method, the supply air fan receives a set-point for the differential pressure that needs to be fulfilled, and the flap angle of a motor flap control and the air volume flow that needs to be supplied in each zone. Furthermore, this solution is developed in a way that it can be easily implemented and integrated into building management applications without additional instruments or setups. Thus, the pressure drop within duct segments is not measured directly but calculated with a model based on the system schematics. This implies that the pressure drop model needs to be developed for each duct system and building topology. Figure 4 illustrates the CO_2 reset strategy.

Two models work together, the pressure drop model of the system and the CO_2 zone model. The first one calculates the current air volume flow into the zone at any moment and adjusts the power of the fan required to fulfil the demand. The second one calculates the CO_2 level of the zone depending on the incoming air flow, the number of occupants and the geometry of the zone. In general terms, the supply air fan receives a pressure set-point and update the current pressure to the dynamic model. Afterwards, the motor flaps receive the angle set points. Again, the CO_2 level in the zone and the flow volume in each terminal are measured together with the pressure and volume flow at the supply air fan to calculate the new set points.

Zucker et al. validated their approach by comparing their model implemented in MATLAB™ and measuring data collected from a ventilation system in a passive house office building in Vienna, Austria. The estimated investment cost saving of motor flaps compared to VAV controller, is around 55% and the new approach improves the energy efficiency by 5%. However, this approach has a significant modelling effort that does not compensate for the savings thus is not practical for implementation. It was suggested to redefine the modelling strategies by using Building Information Model (BIM) and implementing a grey box modelling to reduce the complexity of the models.

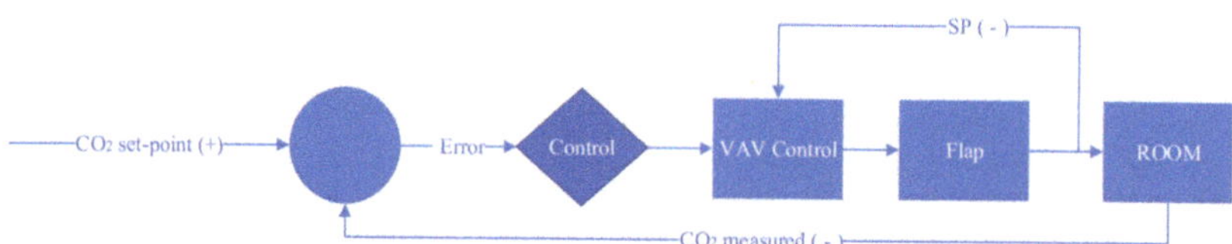

Figure 4. CO_2 reset control strategy [15].

3.3. Fault Tolerant Control Strategies

As it was mentioned earlier, VAV boxes are versatile and flexible devices used to control the airflow and deliver the adequate amount of air in each conditioned space. However, these elements are highly complex, expensive and more importantly prone to failures. These failures need to be identified, isolated and corrected because they have an adverse effect in both, thermal comfort and energy consumption [34]. For this reason, researchers have been focused on the development of fault-adaptive control strategies which consists of two sequential subsystems. The first one provides information about the occurred fault and the second one adapts the control to the faulty operation. As shown by Darure et al. [19], most of the fault-adaptive control strategies are designed to recover the system and the control performance at the cost of higher energy consumption. Instead, they integrate explicitly the energy consumption of the building as an indicator to be minimised under a faulty operation. First, a fault detection and diagnosis module identifies and isolates the damper failure and then, a fault-adaptive control lock-in the issue and makes possible to reduce the energy consumption of the building within desired thermal comfort levels. This solution is tested in a building in the city of Nancy, France.

3.4. Room Pressure-Based Control Strategies

In 2011 a new control strategy for ventilation and conditioning systems was invented by Albert Bauer and patented by Bosch GmbH [27]. It is claimed that this invention allows a cost-efficient and flexible air stream control for improving the comfort and the zone climate. This technology results as an alternative to the volume stream controllers used in conventional systems that, in general, are cost and maintenance intensive and it replaces them by cooperating climate and differential pressure controllers and throttle flaps. It is claimed that, the device controls the inlet and exhaust fans, and the inlet and exhaust throttle flaps in order to maintain the desired levels of zone pressure, room temperature, CO_2. These variables become direct control parameters which allow the system to work with the minimum inlet and exhaust fan power so that the energy is used efficiently, while each zone is conditioned independently as desired within a $\pm 2\,°C$ setpoint.

The system is composed of (1) an inlet and exhaust fan in central inlet and exhaust channels; (2) inlet and exhaust throttles in secondary channels that connect each zone with the central channels; (3) climate sensors in each zone to be conditioned (e.g., temperature, humidity and CO_2 sensors); (4) pressure sensors in each zone to be conditioned. Additionally, there are different controllers to adjust the throttle openings and the AHU fan power as follows:

- Climate controls:

  - First climate controller: controls each zone's input throttle flaps opening cross section to ensure the desired zone climate.
  - Second climate controller: adjust the input fan requirements to ensure enough pressure is available to provide the necessary airflow in all zones.

- Pressure controls:

  - First a fine differential pressure sensor (with pressure piping to indoor & outdoor) and controller: controls each zone's exhaust throttle flaps opening cross section to ensure the desired zone pressure.
  - Second pressure controller: adjust the exhaust fan power required to maintain the desired pressure in all the zones.
  - Third pressure controller: controls each zone's input throttle flaps opening cross section to ensure the desired zone pressure.
  - Fourth pressure controller: controls the inlet AHU fan power in case that the exhaust fan power is not enough to maintain the zone pressure and climate levels in all the zones.

This controller allows maintaining the duct static pressure at minimum level so that all the zones are supplied with the required amount of fresh air. Additionally, it allows to define and control the pressure within each zone. In this way, it is possible to produce a pressure differential between the zone and the atmosphere. This control system needs constant feedback from number of sensors that are connected via BMS, however, the digital infrastructure of BMS does not allow such heavy load of data transition.

More recently, the advancement in the telecommunication technologies has created the new era of the Internet of Things (IoT). The IoT based sensing technologies have penetrated into the HVAC operation and control systems, creating smart control devices for the ventilation and conditioning the building.

A revolutionary approach called "Atomic Air" was invented by Saxby [35] to create superior indoor atmospheres and is claimed to have mechanical energy savings along with superior IAQ. The system allows a cost-efficient installation for creating superior indoor for each zone [35]. This invention utilises a unique control algorithm based on gas laws while utilizing artificial intelligence to finely tune and control the HVAC plant. The algorithm creates a stochastic molecular motion via high-velocity molecular collisions within the space, this creates a completely diffused atmosphere. The solution controls the energy input (heat/coolth, volume and mass), supply and exhaust fans independently, and the supply and exhaust dampers in order to maintain individual room atmospheres, temperature, CO_2 and humidity (Figure 5). These variables become direct control parameters which allow the system to work with the minimum energy input (heat/coolth), supply and exhaust fan power so that the mechanical energy is used efficiently and assisted by the utilisation of the kinetic energy exchange via the molecules. This efficiency means the system no longer has to compensate for convection (via large air volumes) and can use pure thermodynamics to mitigate excess heat or coolth. The overall air treatment volume can be reduced very significantly while not only maintaining but controlling IAQ to programmable and high standards.

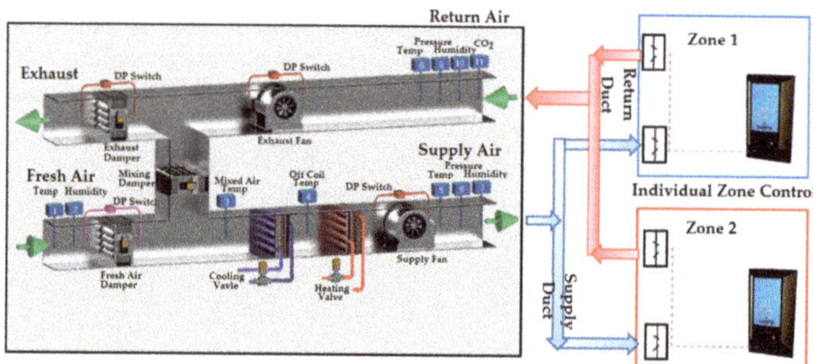

Figure 5. Schematic diagram of the Atomic Air system.

4. Discussion

In previous section the novel and revolutionary strategies to balance the air flow in multi-zone air conditioning systems were reviewed. These solutions are then categorised depending on the device used as terminal: On one hand the VAV boxes are used to measure pressure and air flow as well as to control the air flow; On the other hand, motor flaps are used to control the air flow supply and alternative strategies are developed to measure or estimate the required air flow to be supplied in each zone. Both methods are intended to minimise the air flow that is used by the ventilation system so that the energy consumption is improved. However, as it was later discussed, it is not only about thermal comfort but IAQ. Consequently, this review included technologies that are being developed to achieve ideal IAQ levels or that already implemented a control parameter related to this.

VAV boxes are terminal units equipped with a range of sensors to measure parameters such as air flow, pressure, temperature, etc; dampers and complex electronic components. As a result, they can be used within different types of control strategies, they are versatile and adaptable to different requirements and conditions. Nonetheless, they are expensive, which make them suitable only for medium and large projects, and susceptible to failures, which affects the performance.

Then, the paper presents novel solutions such as "Atomic Air" to balance the air flow using motor flaps instead of VAV boxes. This reduces the investment cost and the failure risk since they are simple devices. On the contrary, without airflow measurement equipment the control strategies must find models or relations with other parameters to control the air flow by adjusting the angle of the flaps. Some new technologies have taken an entirely new approach to the problem looking firstly at the atmosphere and how molecules react in certain circumstances. This combined with real-time plant and damper control has yielded system accuracy to setpoint within 0.4 °C building wide with a fully controlled IAQ and energy savings up to 70%. The result is an effective system that reduces the air volume processed at plant by typically 30–70%, but with-in zone delivery (via molecular expansion) to more than suffice the real-time requirement of IAQ and room condition.

Table 1 summarises the key points of each strategy presented. As shown, first technologies are based on resets of the duct static pressure using VAV boxes. As they do not consider an IAQ parameter the improvement of the energy consumption that they achieved (compared to constant static pressure strategies) could not guarantee the IAQ required. However, Rahnama et al. [23] have started to include CO_2 as demand parameter and the next chapter of their studied will include this within their approach.

Table 1. Control strategies—Summary.

Device	Author	Year	Country	Tech.	Validation	Findings
VAV Box	Koulani et al. [30]	2014	Denmark	Static Pressure Reset (SPR)	Simulation Experiment	Energy consumption reduced by 14% compared to CSP. Control is integrated with BMS. No IAQ parameters involved in the control strategy.
	Walaszczyk and Cichón [18]	2017	Poland	Static Pressure Reset (SPR)	Simulation	Trim & Respond method has a better performance than the PID control. No IAQ parameters involved in the control strategy. Control based on duct pressure instead of damper opening.
	Rahnama et al. [23]	2017	Denmark	Static Pressure Reset (SPR)	Simulation/Experiment	No IAQ parameters involved in the control strategy but planned for future work. Energy consumption reduced in at least 21%.
	Lin and Lau [24]	2014	USA	CO_2 reset	Simulation	CO_2 levels involved in the control strategy for the outdoor air flow. Still requires design parameters for defining occupancy levels. Reduction of outdoor air flow of 14.6%. Monetary savings are estimated between 0.3% and 11%.
	Liu and Lau [25]	2015	USA	CO_2 reset	Simulation	CO_2 levels involved in the control strategy for the outdoor air flow and the zone air-flow supply. System efficiency involved in the control strategy. Reduction of outdoor airflow of approximately 45%. Monetary savings are estimated between 25% and 45% approximately.
	Kim et al. [22]	2017	Korea	CO_2 reset	Simulation	Control both recirculation air flow and outdoor airflow depending on zone CO_2 level. Thermal comfort and IAQ levels achieved. Energy consumption reduction of 20% compared to fixed air flow systems.
Motor Flaps	Robert Bosch GMBH [27]	2015	Germany	Zone pressure-based	Patent	Zone pressure and climate parameters (i.e., Temperature and CO_2) controls supply and exhaust fans and flaps. An overpressure or under pressure could be set for each zone. A pressure model is used to estimate the supply airflow of each zone and adjust the fan power.
	Zucker et al. [15]	2017	Austria	CO_2 and Static Pressure reset based	Simulation	A CO_2 model is used to estimate zone requirement of air flow. Models developed are too complicated and different for each scenario so that this solution is not practical. Investment cost savings are estimated in 55%. Improvement in the energy efficiency of 5%.
	Jing et al. [33]	2019	China & Singapore	Improved SPR	Experimental	A model-based, improved SPR strategy is used to well-balance the ventilation system. A damper position control method and an optimal static pressure set-point selection method guarantee the system is energy efficient and, at the same time, the required airflows are achieved. This strategy achieves energy savings of 21% relative to traditional SPR. This methodology is complex so that for large-scale, ducted networks, the model of the systems needs to be simplified.
Smart VAV-T with motor Flaps	Saxby and Tuck [35]	2016	UK	Gas Law algorithm	Patent Pending	System & Controls solution; based on gas law equations with kinetic energy transfer. SMART system with IoT and real-time monitoring, analysis, control, reaction, reporting, and alarming. The Atomic Air algorithm calculates in real-time the required energy demand, fresh/re-circ air mixing, supply-return fan (independently in 1% increments) based on (individual) in zone molecular expansion ratios, heat loads and air quality. Resulting in near zero stratification (via stochastic molecular motion). Investment cost savings are typically 30–60% Improvement in energy savings 40–70% Controlling the CO_2 level 100%

In connection with the CO_2-based reset control strategies, they improved the energy consumption as well as the IAQ. However, they still rely on design setpoints to adjust the air flow. By scheduling the expected occupancy in the zones together with real-time CO_2 measurements, the air flow is adjusted. These alternatives have an important potential since they could be responding to predefined conditions and not the actual demand.

Furthermore, it can be noted that many studies focused on solutions to improve the energy consumption while achieving ideal IAQ and thermal comfort levels using VAV boxes to control and measure the airflow. Nevertheless, it has been demonstrated that it is possible to use motor flaps to reduce complexity and costs. For example, Zucker et al. [15] developed a motor flap technology based on CO_2 reset controls couple with the pressure. However, this solution is impractical due to the complexity of the model. To be implemented, the models developed to estimate the required airflow need to be simplified. Instead, Bosch's novel approach [27] controls the airflow supplied in each zone depending on climate sensors such as temperature and CO_2 sensors together with a zone to outdoor differential pressure sensor. However, it can cause issues in tall buildings, at altitude and with vapour pressure.

5. Conclusions

This paper presents a review concerning novel technologies to minimise the energy consumption of ventilation systems. Two categories were investigated, first VAV systems that control the airflow using VAV boxes and then, solutions that use motor flaps. While many studies have focused on the first one, developing static pressure rests and CO_2 reset models is considered to be the novel approach that is fast growing. There are many opportunities to further exploit the potential of the presented solutions by ensuring that the supply airflow is answering the real-time demand condition of the zones. An alternative approach in the first category is a hybrid system that utilise VAV combined with the motor flaps. This has found to have the best outcome in terms of IAQ and energy saving. There is a limited study presented on the hybrid approach. Their performance could be further improved by studying in depth the effect of the overpressure on the IAQ. These systems use complex mathematics to control the VAV systems. Therefore, they have been designed to offer a plug and play solution, that can be added as a new installation or can use in an already installed VAV or motor flaps to improve the system performance. Finally, tailored computer modelling need to be developed to facilitate the implementation of these technologies and how they can be improved through DC fans, compressor and chilled water technologies during the HVAC system design.

Author Contributions: All the authors contributed to publishing this paper. B.R. and J.M.Z. prepared and edited the manuscript. B.S. and R.T. revised and reviewed the manuscript. M.S. provides technical data and drawings as well as review and revision.

Funding: This research received no external funding.

Conflicts of Interest: The authors declare no conflict of interest.

References

1. IEA. *World Energy Outlook Special Report*; International Energy Agency: Paris, France, 2015.
2. IPCC. *Climate Change 2014: Mitigation of Climate Change*; Intergovernmental Panel on Climate Change: Cambridge, UK; New York, NY, USA, 2014.
3. United Nations. *The Paris Agreement on Climate Chang*; United Nations: Rome, Italy, 2015.
4. Anderson, J.E.; Wulfhorst, G.; Lang, W. Energy analysis of the built environment—A review and outlook. *Renew. Sustain. Energy Rev.* **2015**, *44*, 149–158. [CrossRef]
5. Liu, D.; Zhao, F.-Y.; Tang, G.-F. Active low-grade energy recovery potential for building energy conservation. *Renew. Sustain. Energy Rev.* **2010**, *14*, 2736–2747. [CrossRef]
6. Wang, Y.; Kuckelkorn, J.; Liu, Y. A state of art review on methodologies for control strategies in low energy buildings in the period from 2006 to 2016. *Energy Build.* **2017**, *147*, 27–40. [CrossRef]

7.   Afram, A.; Janabi-Sharifi, F.; Fung, A.S.; Raahemifar, K. Artificial neural network (ann) based model predictive control (mpc) and optimization of hvac systems: A state of the art review and case study of a residential hvac system. *Energy Build.* **2017**, *141*, 96–113. [CrossRef]

8.   Pérez-Lombard, L.; Ortiz, J.; Pout, C. A review on buildings energy consumption information. *Energy Build.* **2008**, *40*, 394–398. [CrossRef]

9.   Chenari, B.; Dias Carrilho, J.; Gameiro da Silva, M. Towards sustainable, energy-efficient and healthy ventilation strategies in buildings: A review. *Renew. Sustain. Energy Rev.* **2016**, *59*, 1426–1447. [CrossRef]

10.   Okochi, G.S.; Yao, Y. A review of recent developments and technological advancements of variable-air-volume (vav) air-conditioning systems. *Renew. Sustain. Energy Rev.* **2016**, *59*, 784–817. [CrossRef]

11.   Al horr, Y.; Arif, M.; Katafygiotou, M.; Mazroei, A.; Kaushik, A.; Elsarrag, E. Impact of indoor environmental quality on occupant well-being and comfort: A review of the literature. *Int. J. Sustain. Built Environ.* **2016**, *5*, 1–11. [CrossRef]

12.   Hormigos-Jimenez, S.; Padilla-Marcos, M.Á.; Meiss, A.; Gonzalez-Lezcano, R.A.; Feijó-Muñoz, J. Ventilation rate determination method for residential buildings according to tvoc emissions from building materials. *Build. Environ.* **2017**, *123*, 555–563. [CrossRef]

13.   Steinemann, A.; Wargocki, P.; Rismanchi, B. Ten questions concerning green buildings and indoor air quality. *Build. Environ.* **2017**, *112*, 351–358. [CrossRef]

14.   Chen, H.; Cai, W.; Chen, C. Fan-independent air balancing method based on computation model of air duct system. *Build. Environ.* **2016**, *105*, 295–306. [CrossRef]

15.   Zucker, G.; Sporr, A.; Garrido-Marijuan, A.; Ferhatbegovic, T.; Hofmann, R. A ventilation system controller based on pressure-drop and co2 models. *Energy Build.* **2017**, *155*, 378–389. [CrossRef]

16.   Shim, G.; Song, L.; Wang, G. Comparison of different fan control strategies on a variable air volume systems through simulations and experiments. *Build. Environ.* **2014**, *72*, 212–222. [CrossRef]

17.   Pang, X.; Piette, M.A.; Zhou, N. Characterizing variations in variable air volume system controls. *Energy Build.* **2017**, *135*, 166–175. [CrossRef]

18.   Walaszczyk, J.; Cichoń, A. Impact of the duct static pressure reset control strategy on the energy consumption by the hvac system. In Proceedings of the 9th Conference on Interdisciplinary Problems in Environmental Protection and Engineering EKO-DOK 2017, Boguszow-Gorce, Poland, 23–25 April 2017; EDP Sciences: Paris, France, 2017; p. 9.

19.   Darure, T.; Yamé, J.J.; Hamelin, F. Fault-adaptive control of vav damper stuck in a multizone building. In Proceedings of the 2016 3rd Conference on Control and Fault-Tolerant Systems (SysTol), Barcelona, Catalonia, Spain, 7–9 September 2016; pp. 170–176.

20.   Hartadi, S. Schematic Diagram of Air Conditioning System. Vol. 2040 × 1152, p Schematic Diagram of a Typical Variable Air Volume (VAV) System. 2018. Available online: http://www.goethes-farbenlehre.com/ (accessed on 10 December 2018).

21.   Westphalen, D.; Koszalinski, S. *Energy Consumption Characteristics of Commercial Building Hvac Systems*; U.S. Department of Energy: Washington, DC, USA, 1999.

22.   Kim, H.-J.; Cho, Y.H. A study on a control method with a ventilation requirement of a vav system in multi-zone. *Sustainability* **2017**, *9*, 2066. [CrossRef]

23.   Rahnama, S.; Afshari, A.; Bergsøe, N.C.; Sadrizadeh, S. Experimental study of the pressure reset control strategy for energy-efficient fan operation: Part 1: Variable air volume ventilation system with dampers. *Energy Build.* **2017**, *139*, 72–77. [CrossRef]

24.   Lin, X.; Lau, J. Demand controlled ventilation for multiple zone hvac systems: Co2-based dynamic reset (rp 1547). *HVAC&R Res.* **2014**, *20*, 875–888.

25.   Lin, X.; Lau, J. Demand-controlled ventilation for multiple-zone hvac systems—part 2: Co2-based dynamic reset with zone primary airflow minimum set-point reset (rp-1547). *Sci. Technol. Built Environ.* **2015**, *21*, 1100–1108. [CrossRef]

26.   Kaam, S.; Raftery, P.; Cheng, H.; Paliaga, G. Time-averaged ventilation for optimized control of variable-air-volume systems. *Energy Build.* **2017**, *139*, 465–475. [CrossRef]

27.   Bauer, A. Control Device for Ventilation and Air Conditioning Systems. Google Patents US20110300790A1, 2011.

28.   Liu, G.; Liu, M. Development of simplified in-situ fan curve measurement method using the manufacturers fan curve. *Build. Environ.* **2012**, *48*, 77–83. [CrossRef]

29. Zhang, J.; Li, X.; Zhao, T.; Dai, W. Experimental study on a novel fuzzy control method for static pressure reset based on the maximum damper position feedback. *Energy Build.* **2015**, *108*, 215–222. [CrossRef]

30. Koulani, C.; Hviid, C.; Terkildsen, S. Optimized damper control of pressure and airflow in ventilation systems. In Proceedings of the 10th Nordic Symposium on Building Physics, Lund, Sweden, 15–19 June 2014; pp. 822–829.

31. Rahnama, S.; Afshari, A.; Bergsøe, N.C.; Sadrizadeh, S.; Hultmark, G. Experimental study of the pressure reset control strategy for energy-efficient fan operation—Part 2: Variable air volume ventilation system with decentralized fans. *Energy Build.* **2018**, *172*, 249–256. [CrossRef]

32. Tukur, A.; Hallinan, K.P. Statistically informed static pressure control in multiple-zone vav systems. *Energy Build.* **2017**, *135*, 244–252. [CrossRef]

33. Jing, G.; Cai, W.; Zhang, X.; Cui, C.; Yin, X.; Xian, H. Modeling, air balancing and optimal pressure set-point selection for the ventilation system with minimized energy consumption. *App. Energy* **2019**, *236*, 574–589. [CrossRef]

34. Liang, Y.; Meng, Q.; Chang, S. Fault diagnosis and energy consumption analysis for variable air volume air conditioning system: A case study. *Procedia Eng.* **2017**, *205*, 834–841. [CrossRef]

35. Saxby, B.; Tuck, R. Atomic Air, System & Controls Solution; Based on Gas Law Equations with Unique Kinetic Energy Transfer. Available online: https://www.oairo.me/ (accessed on 1 April 2018).

Article

Assessing the Demand Side Management Potential and the Energy Flexibility of Heat Pumps in Buildings

Alessia Arteconi [1,*] and Fabio Polonara [2]

1   Università eCampus, via Isimbardi 10, 22060 Novedrate (CO), Italy
2   Dipartimento di Ingegneria Industriale e Scienze Matematiche, Università Politecnica delle Marche,
     via brecce bianche 1, 60131 Ancona, Italy; f.polonara@univpm.it
*   Correspondence: alessia.arteconi@uniecampus.it; Tel.: +39-071-220-4432

Received: 29 June 2018; Accepted: 13 July 2018; Published: 14 July 2018

Abstract: The energy demand in buildings represents a considerable share of the overall energy use. Given the significance and acknowledged flexibility of thermostatically controlled loads, they represent an interesting option for the implementation of demand side management (DSM) strategies. In this paper, an overview of the possible DSM applications in the field of air conditioning and heat pumps is provided. In particular, the focus is on the heat pump sector. Three case studies are analyzed in order to assess the energy flexibility provided by DSM technologies classified as energy efficient devices, energy storage systems, and demand response programs. The load shifting potential, in terms of power and time, is evaluated by varying the system configuration. Main findings show that energy efficient devices perform strategic conservation and peak shaving strategies, energy storage systems perform load shifting, while demand response programs perform peak shaving and valley filling strategies.

Keywords: demand side management (DSM); energy efficiency; energy storage; demand response (DR); flexibility

---

1. Introduction

Energy consumption in the residential sector represents about 40% of the total energy use both in Europe and in the US [1,2]. In particular, heating and cooling in buildings have a high share of the overall energy use. Air conditioners are responsible on average for around 5% of global electricity consumption, with a variable percentage country by country (e.g., 14% in the US and 40% in India) [3], and such share is expected to grow due to the increasing refrigeration demand also due to global warming. The relevance of heat pumps is also increasing; in 2015 about 800,000 units were sold in Europe and a growing trend in sales is estimated for the coming years [4]. Nevertheless, buildings are considered a resource in power systems thanks to the high energy flexibility they promise. Indeed, in buildings there are several deferrable loads (e.g., laundry machines and dish washers) and thermostatically controlled loads (TCL), such as heat pumps, refrigerators, and air conditioners. The latter technologies, especially, contain various forms of storage, which can be used to alter the electric load without affecting the quality of the energy service [5]. The energy flexibility provided by buildings is paramount to mitigate the upcoming challenges of future power systems, and its exact definition and quantification has a central role, as stated by the working group of Annex 67 about energy flexible buildings [6]. Furthermore, the EU's Energy Performance of Buildings Directive [7] pushes towards new and more performing buildings—nearly zero energy buildings (nZEB)—where energy efficiency and energy flexibility are essential to achieve the required performance targets.

Given this premise, the sector of heating and cooling in buildings is promising for the application of demand side management (DSM) strategies aimed at modifying the final user's electricity demand on the basis of electricity grid needs. The relevance of DSM is related to the growing share of renewable energy sources (RES) in the generation mix and consequently to the necessity of integrating them and of adapting the energy demand to their intermittent and unpredictable production. Besides this aspect, DSM can produce several other advantages in the electric power system, which can be summarized as [8]: reduced need for increasing the power generation capacity; higher operational efficiency in production, transmission, and distribution of electric power; and lower electricity costs. A DSM strategy can have different purposes, such as peak shaving, valley filling, load shifting, energy conservation, and strategic load growth [9].

DSM technologies can be used to activate the energy flexibility of buildings. They can be divided into three main categories: (i) energy-efficient end-use devices; (ii) additional equipment, systems, and controls to enable load shaping (e.g., energy storage); and (iii) communication systems between end-users and external parties, for example, demand response (DR) programs [9]. While the first point concerns energy conservation by means of devices using less energy, the other two points deal with systems aimed at shifting the final user's demand: thermal energy storage systems can be used to store surplus energy to be released for later use, whereas demand response achieves changes in final users' load by means of price signals from the grid.

The purpose of this paper is to highlight how DSM technologies for air conditioning and heat pumps can unlock the energy flexibility in buildings and to show, consequently, which kind of DSM strategy can be fulfilled through them. A state of the art review is presented to illustrate the existing trend in the implementation of the above mentioned DSM categories (Section 2), that is, (i) energy efficiency, (ii) energy storage systems, and (iii) demand response programs. In this way, a holistic overview of the potential for managing the energy demand related to cooling or heating loads is provided. Considering the relevance in terms of share on global energy use and units sold worldwide, and the increasing importance of electric energy among the final uses, a particular focus is on the heat pump sector. Three different case studies for each of the considered DSM categories are analyzed and the consequent energy flexibility is quantified by means of flexibility indicators available in the literature and illustrated in Section 3. Indeed, it is shown that several indicators for energy flexibility quantification have been introduced; here the main novelty lays in the application of the most meaningful to compare the flexibility potential of different DSM technologies. By means of these case studies, an in-depth analysis on more efficient devices, energy storage systems, and DR actions is performed and their effect in terms of modification of energy demand load shape is discussed (Section 4). Some conclusions about unlocking energy flexibility are drawn on the basis of the obtained results (Section 5).

2. State of the Art Review of DSM Technologies for Air Conditioning and Heat Pumps

In this section, the process of how demand side management can be realized in the HVAC (heating ventilation and air conditioning) sector is discussed. Chillers and heat pumps, as said, belong in the category of thermostatically controlled loads: they are aimed at keeping the operative temperature in a given range and they allow the shift of thermal loads produced by electricity conversion, thanks to inherent or external thermal energy storages. The results of the literature review are divided on the basis of the above-mentioned categorization of DSM technologies: energy efficiency, energy storage, and demand response. Main findings are summarized in Table 1 and are discussed in detail in the following sections.

Table 1. Summary of the demand side management (DSM) technologies implemented in the heating ventilation and air conditioning (HVAC) sector.

Category	Technology
Energy efficiency	Natural refrigerants Optimal design Not-in-kind technologies Optimal control strategies
Storage	Cold Thermal Energy Storage/Thermal Energy Storage Thermally Activated Buildings Phase Change Material
Demand Response	Ancillary services (frequency and voltage control, congestion management, spinning and non-spinning reserves) Renewable Energy Sources integration Cost reduction

2.1. Energy Efficiency

Energy efficiency in reverse cycle devices (i.e., heat pumps and chillers) can be achieved in several ways. It is possible to act on: (i) the choice of a proper and environmental friendly refrigerant; (ii) more efficient design of components (variable speed compressors, variable refrigerant flow, ejectors); (iii) application of not-in-kind technologies; and (iv) optimization of the overall HVAC system and control strategies.

In the literature there are several studies that deal with the above mentioned topics. In the following, some of them are reported:

- The use of natural refrigerants is gaining more and more interest. Harby [10] focuses, for example, on the problems (environmental impacts and energy use) associated with the use of halogenated refrigerants and the possibility to replace them with hydrocarbons (HC). The author provides a guide to understand the current status, replacement strategy of conventional refrigerants, and possible drawbacks in using HC. Bolaji and Huan [11] review existing natural refrigerants (HC, ammonia, CO_2, water) and their areas of application in refrigeration and air-conditioning systems. They show the potential of natural refrigerants to replace halogenated refrigerants. Nawaz et al. [12] propose propane and isobutene as natural refrigerants for heat pump water heaters, while Zajacs et al. [13] analyze the possibility of using ammonia in a heat pump for space heating and domestic hot water production.

- Among the other technologies, variable capacity compressors are considered as an effective way to achieve energy efficiency for air-conditioning chillers and heat pumps. Krishnamoorthy et al. [14] present the cooling operation of a variable-capacity/variable-fan-speed heat pump and they show its potential to reduce residential energy use, highlighting also the importance of location and insulation level of the duct system. Park et al. [15] evaluate the performance of a variable refrigerant flow (VRF) air conditioning (AC), that is, an air-conditioning system able to control the amount of refrigerant flowing to the multiple evaporators (indoor units). They assess energy use and comfort of occupants for a Korean building and obtained results confirming the advantages of a VRF system. Aynur et al. [16] compare a VRF AC system with a variable air volume (VAV) AC and assess a potential energy saving between 27% and 58% depending on the system configuration, and indoor and outdoor conditions. Özahi et al. [17] show for a real case that variable refrigerant flow systems can produce a 44% cost saving compared to traditional HVAC systems. Among the ways of improving the performance of a vapor compression refrigeration cycle, using an ejector as an expansion device is one of the possibilities [18]. The ejector is a machine without movable organs that can be used as a compressor and as a pump to achieve a pressure rise of a given fluid through supply of another fluid with the same or different nature. Wang et al. [19] describe a hybrid ejector air-conditioning system and its optimization. The proposed system has the compressor installed between the outlet of evaporator and ejector secondary flow inlet. The optimization of such a system leads to a 7% energy use reduction. Boccardi et al. [20], instead, evaluate the performance improvement due to an ejector in a HVAC system working with CO_2 as refrigerant. The optimal configuration of the system is determined.

- Not-in-kind technologies category includes the following technologies for domestic applications: thermoacoustic refrigeration, thermoelectric refrigeration, thermotunneling, magnetic refrigeration, Stirling cycle refrigeration, pulse tube refrigeration, Malone cycle refrigeration, absorption refrigeration, adsorption refrigeration, and compressor driven metal hydride heat pumps [21]. Among all the others, magnetic refrigeration shows fast developments in new materials and system architecture, and it is considered very promising as an alternative to some conventional cooling, refrigeration, and air-conditioning technologies [22].

- Energy recovery is of paramount importance to increase the efficiency of an HVAC and it can be performed, for example, by means of an enthalpy wheel or heat exchangers, whose performance can be enhanced by the inclusion of phase change material (PCM) [23]. Du et al. [24], instead, analyze different control strategies for achieving energy savings with a HVAC system in an airport. They make a comparison on the basis of an exergy based method. Ascione et al. [25] present a methodology for the optimization of a building-HVAC system by employing a model predictive control (MPC) strategy for heating and cooling. The authors show a primary energy saving around 35.4 kWh/m^2 for a residential building in Southern Italy. Indeed, several studies have shown that by adding a supervisory MPC controller to HVAC systems could result in energy use and operating cost reductions from 7% to more than 50% [26–28].

2.2. Energy Storage

The relevance of coupling reverse cycle devices with thermal energy storage in order to modify the final user's demand and match it with the power system production has been demonstrated in the literature [29].

A cold thermal energy storage (CTES) can be coupled to a chiller for air conditioning. It can be a thermally stratified chilled water storage (sensible TES) or an ice storage (latent TES) [30]. A TES can convey several benefits: improving the refrigeration equipment efficiency, reducing the installed capacity, increasing the operational flexibility, and reducing energy costs, because cold is produced during off-peak periods and used during on-peak periods. Typical operational strategies of such systems are full storage and partial storage strategies, as shown in Figure 1 [31].

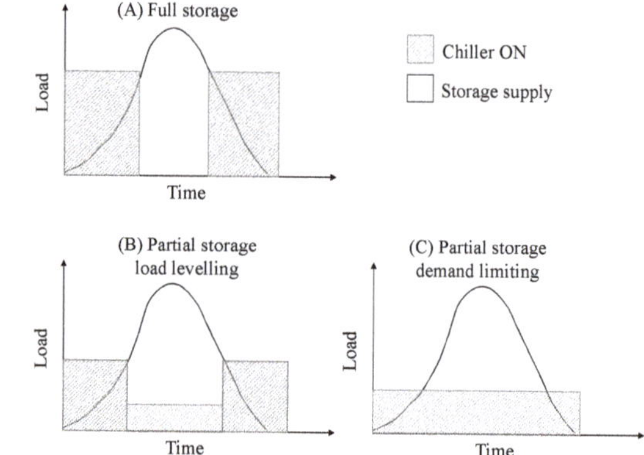

Figure 1. Cold thermal energy storage (CTES) operational strategies [29,30]. (**A**) A full storage strategy shifts the entire peak cooling load to off-peak hours. In a partial storage strategy, the load is partially supplied by the thermal storage and partially by the refrigeration unit. The chiller can be designed to operate as (**B**) load levelling or (**C**) demand limiting.

Partial storage is the most used configuration because of lower initial capital costs. It saves about 40–60% of peak cooling electricity demand. Full storage is interesting for short peak periods with very high costs for electricity production. The peak electricity demand can be reduced by up to 80–90% [32]. Similarly, thermal energy storage systems can be coupled with heat pumps for implementing demand side management strategies [33], such as peak shaving or night load shifting. These kinds of applications can lead to energy bill reductions, even if generally the energy use increases because of thermal losses in the storage systems.

Rather than the use of external storage tanks, also much more effective in buildings, there is the use of the building envelope—thermal mass as passive or active energy storage [34]. Floor heating or TABs (thermally activated buildings) are active thermal storage systems [35]: the floor is a large, low temperature radiating surface while the concrete acts like a thermal storage heated by air, water, or directly by electric resistances. The building envelope, instead, is passive thermal storage, easily exploited by means of variable inside temperature set-points, which can be adjusted in a certain range in order to maintain the internal comfort and, at the same time, provide useful flexibility to the power system [36]. Furthermore, PCM can be integrated within the building components and used passively, for pre-cooling or pre-heating and for reducing the thermal load request, or alternatively used actively, by charging and discharging their energy content on the basis of a given strategy.

2.3. Demand Response

Among the other devices, reversible heat pumps (for heating and cooling) are recognized as systems to be effectively used to implement demand response strategies in the smart grid. Demand response strategies are mainly aimed at three different purposes: (i) provision of ancillary services for the power grid, (ii) renewable energy sources integration, and (iii) energy bill reduction for the final user [37].

- Ancillary services consist of frequency and voltage control, congestion management, and provision of spinning and non-spinning reserve [37]. In voltage control applications, heat pump load is modified to maintain the distribution network voltage within the allowed limits: heat pump (HP) active power demand is decreased in times of local under-voltage and increased in times of over-voltage. The latter case can happen, for example, in presence of photovoltaic (PV) panels, feeding electricity in the grid. In a similar manner, frequency control can be realized to ensure that the grid frequency stays within a specific range of the nominal frequency. An existing example of application is the virtual energy storage network, named "tiko", which connects more than 10,000 electric heating devices (i.e., heat pumps and water heaters) throughout Switzerland [38]. As far as congestion management in the distribution grid is concerned, optimized scheduling and power rating of heat pumps are paramount in order to avoid limitations in the transformer and line capacity [37,39]. Eventually heat pumps can serve as spinning and non-spinning reserves used by the transmission system operator (TSO) [37,40], that is, as reserve capacity that can be regulated downward or upward to match demand and supply. Different strategies can be implemented with these reserves, such as peak clipping or valley filling. In References [41–43], incentive based DR programs are considered to provide ancillary services to the power system by means of active demand and their influence on the real power markets is assessed.
- Another important role played by heat pumps and the flexibility they provide is related to the possibility to integrate renewable energy sources both at building level and at power system level. The main RES considered are wind and PV. Bruninx [40] shows the impact on wind utilization factor (WUF) produced by heat pumps adhering to a DR program: on average, the WUF increases from about 75% in the reference case to 84% when heat pumps are used as operational reserves. This increase of WUF is the result of shifting demand to periods of excess wind power generation and of increasing the indoor temperature in order to allow the DR adherent heating systems to provide upward reserves. Baetens et al. [44] demonstrate the positive role of heat pumps

to reduce curtailment of electricity from PV panels. Arteconi et al. [45] show how heat pumps coupled with thermal storage can increase the self-consumption of PV electricity production in an industrial building. In this study, the tank operative conditions are analyzed in order to find the best configuration: up to 80% of PV electricity can be used to cover the thermal demand coming from the building.

- Regarding energy bill reduction, electricity price signals are used to influence the final user's consumption pattern. Mainly, there are two types of tariff structures: Time of Use (TOU) tariffs, varying on the period of the day, and Real Time Pricing (RTP), changing with the electricity market price [46]. The heat pump demand can be shifted driven by the price with the intent of reducing electricity costs, improving the system performance, increasing RES integration, or improving power system benefits (i.e., reducing total operational costs). It is evident how the final users and power system interests are strictly related. However, it has been demonstrated that the benefit that a final user can achieve in taking part in a DR program depends also on the number of participants: the operational cost reduction per participant drops when increasing the penetration of DR programs because less flexibility is requested of each user. In more detail, the total cost savings have been assessed at most between about 400 € to 200 € per participant per year for a 5% and 100% DR penetration rate, respectively [47]. In Reference [48], price-controlled energy management of smart homes (SHs) is investigated and it is demonstrated that the benefit for both the generation scheduling problem and for the final users, is a decrease in operational costs. Even in Reference [49], demand response programs are modeled in the unit commitment (UC) problem in order to evaluate achievable operational costs savings.

3. Methods

In the following, three particular DSM technologies, among those above mentioned, were studied. They were chosen on the basis of their widespread and technological relevance. Variable capacity heat pumps were considered as energy efficient devices; thermally stratified water tanks as storage systems; and real time pricing as demand response programs. Firstly, the flexibility assessment procedures available in the literature are listed and the load shifting indicator used here is described in detail. Secondly, the three case studies selected are presented. Eventually, the behavior of the three systems in the presence of the DSM technology is analyzed, and results for the different cases are discussed and compared.

3.1. Flexibility Evaluation

The flexibility provided by heating and cooling systems in buildings is intended as the power consumption deviation of the heating or cooling device from its optimal value (or normal operation) to a new profile aimed at compensating power imbalances in the grid. Such new load patterns can be obtained by using heat buffers where heat or cold can be stored when a surplus of electricity is available and vice-versa. In this context, heat buffers are mainly represented by thermal energy storage and building thermal mass.

There are several studies in the literature that have proposed ways to quantify this flexibility as:

- number of hours the energy use can be delayed or anticipated [50];
- power increase or decrease and time duration of its variation within functional and comfort limits [51];
- energy cost, based on cost functions, which represent how much shifting the energy use costs [52];
- ratio between the maximum change in power to the additional energy use in a certain period (efficiency curves) [53];
- power demand versus priority to be supplied in order to satisfy the operational constraints (priority curves) [54];

- maximum heat that can be stored in the structural storage capacity of a building without violating the thermal comfort [55];
- hourly (h) load shed potential of a device with respect to its rated power consumption (P^{base}) during a DR event (DR^p), as shown in Equation (1) [36]:

$$DR^p = \frac{P_h^{base} - P_h^{LS}}{P_h^{base}} \tag{1}$$

It is evident that different energy flexibility indicators have been proposed and a unique definition does not exist [6]. The most used variables to characterize the flexibility are the amount of power change, duration of the change, rate of change, response time, shifted load and maximal hours of load shifted, and recovery time (which represents when the system is ready to be used again) [56]. The energy flexibility can be evaluated at the design stage, as an intrinsic value related to the building and its heating and cooling system features, or during its operational behavior. The first evaluation is important to have an overall idea of the maximum potential flexibility related to building energy systems, while the second one is useful to assess what happens during operation, especially in case of a specific event, such as a DR control signal sent to the building. In the latter case, the most relevant parameters to take into account during the evaluation are temporal flexibility, power capacity, and energy shifted [57].

In this study, not the normal operation of a system, but the effect of the substitution of a conventional system with another with a higher flexibility potential (i.e., a DSM technology) is considered. The operational behavior is analyzed to evaluate the impact of the DSM technology on the DSM strategy that can be implemented through it. The energy flexibility indicator chosen strictly depends on the purpose of its use. Among others, the number of hours the energy use can be shifted and the hourly load shed potential are indicators highly appropriate to assess the effects of a specific DSM technology. Indeed, from the electricity grid perspective, how the power demand can be modified and for how much time are the most important pieces of information. In the following sections, some case studies where DSM technologies are introduced, are analyzed. The flexibility indicators used are the time shifting and the load shifting potential. The time shifting (t_S) can be assessed by comparing the energy cumulative curves in the case with and without DSM implementation (i.e., the horizontal distance between the curves). Whereas the load shifting potential for the DSM technology, useful when the dynamic operation is evaluated, is calculated similarly to DRp [36] as follows:

$$LS_{DSM,t} = \frac{P_t^{DSM} - P_t^{noDSM}}{P_t^{noDSM}} \tag{2}$$

where $LS_{DSM,t}$ is the load shifting potential for the considered DSM technology at the time step t, P_t^{DSM} is the power consumption at the time step t when the DSM technology is in place, while P_t^{noDSM} is the power consumption at the time step t without any DSM technology. $LS_{DSM,t}$ is positive when the introduction of the DSM technology increases the power load at the considered time step (when P_t^{noDSM} is zero, then $LS_{DSM,t}$ is 100%); it is negative when the DSM technology reduces the power consumption.

3.2. Case Studies

Three different case studies were considered to evaluate the energy flexibility available when a DSM technology was implemented in comparison with reference cases without DSM. In particular, given the relevance of the energy use in the building sector [7] and the increasing market penetration of heat pumps as heating and cooling devices [4], the presented case studies analyzed the implementation of different configurations of heat pump systems in residential buildings. Indeed, it is highly recognized also in Italy the potential of heat pumps as electric heating systems to perform primary energy savings [58]. The energy flexibility is provided by:

1. a HP energy efficient design (variable capacity HP),
2. a thermally stratified water tank,
3. a DR program based on dynamic pricing

and is quantified by means of the above mentioned flexibility indicators. Different types of buildings, heating and cooling systems, and distribution systems were considered in order to make broader the evaluation of the energy flexibility associated with heat pump operation. In particular, the building specifications and generation and distribution characteristics have been chosen to be congruent among them, and with the DSM technology implemented. The buildings were assumed located in a town of central Italy (Ancona). The purpose of the analysis being a qualitative evaluation of the kind of flexibility provided by different DSM technologies, results are shown for a representative winter's day. The inside temperature set-point was set at 20 °C ($\pm$0.5 °C) to be maintained 24 h per day. It was assumed about 30 m^2 per person, which provided an internal gain of 120 W each. A 0.5 ACH (air changes per hour) was modelled. The heat pump performance was obtained from manufacturer's data [59], where the heat pump compressor power consumption was given by varying the evaporator and condenser temperature in relation to the indoor and ambient temperatures. TRNSYS [60] was the tool used for the dynamic simulations. It is widely used for building simulation and it has been thoroughly tested, thus its results can be considered reliable in respect to the issue of model validation, which has a paramount relevance as highlighted in Reference [61]. The simulations time step was 6 min. The case studies are discussed in detail in the following sections. In Table 2 the building properties for each case study are reported; buildings have been represented with a regular shape (parallelepiped). In Table 3 the design parameters of the heating systems are summarized.

Table 2. Specifications of building properties [62,63].

Case	Area m^2	S/V	Windows Area %	U_Wall W/m^{2}K	U_Roof W/m^{2}K	U_Windows W/m^{2}K
A	1000	0.37	10	0.30	0.23	1.4
B	100	0.37	10	0.36	0.28	2.10
C	100	0.37	10	0.36	0.28	2.10

Table 3. Design parameters of the heating system.

Specification	Case A	Case B	Case C
Heating system	GCHP	ASHP	ASHP
Distribution system	floor	radiators	floor
Heating system supply temperature	20–40 °C	40–50 °C	20–40 °C
HP design capacity	8 kW	3 kW	2 kW
COP$_{design}$	4	2.5	3
Tank volume	1000 L	20 L, 100 L, 200 L, 300 L, 1000 L	–

3.2.1. Case A: Energy Efficiency

Case A assessed the flexibility due to an efficient design of the heat pump, that is, a variable capacity heat pump was used. A multi-apartment residential building with a ground coupled heat pump (GCHP) as a heating production system was considered (Figure 2a). It had been modelled in TRNSYS by means of Type 56 for buildings with active layers, while Type 927 for water-to-water heat pumps has been used, acting on the scale factor to simulate the variable capacity, which was coupled to Type 557 for the borefield. Given the novelty of this kind of building in Italy, a new building highly insulated [63] was assumed (see Table 2), having a total floor area of 1000 m^2. Underfloor space heating was used. A 1000 L thermal energy storage was installed to decouple energy production and consumption and to guarantee that space heating was supplied with warm water within a given temperature range (20–40 °C). The heat pump design heating capacity has been assessed at 8 kW, as 70% of the design thermal power (evaluated at the design outside temperature of $-1\ ^\circ$C and inside temperature of 20 $^\circ$C). Indeed, it is common practice in heat pump generation systems with underfloor heating systems to size the heating capacity a bit lower than the design value, given the low occurrence of the design outdoor temperature. In this way, the heat pump is not too oversized for the average operating conditions and a back-up generator can be included for peak power demand. In this exercise, the flexibility provided by the energy efficient design of the heat pump was evaluated: the performance of a variable capacity heat pump, which can reduce its load up to 30% of the design value, was compared to an ON-OFF heat pump. The control logic switches on the heat pump when the ambient temperature is lower than 15 $^\circ$C and the tank temperature goes lower than the upper bound. In presence of the variable capacity heat pump, the capacity was assessed on the basis of the actual building's heat demand. The pump supplying hot water to the distribution system was activated by the indoor temperature set-points.

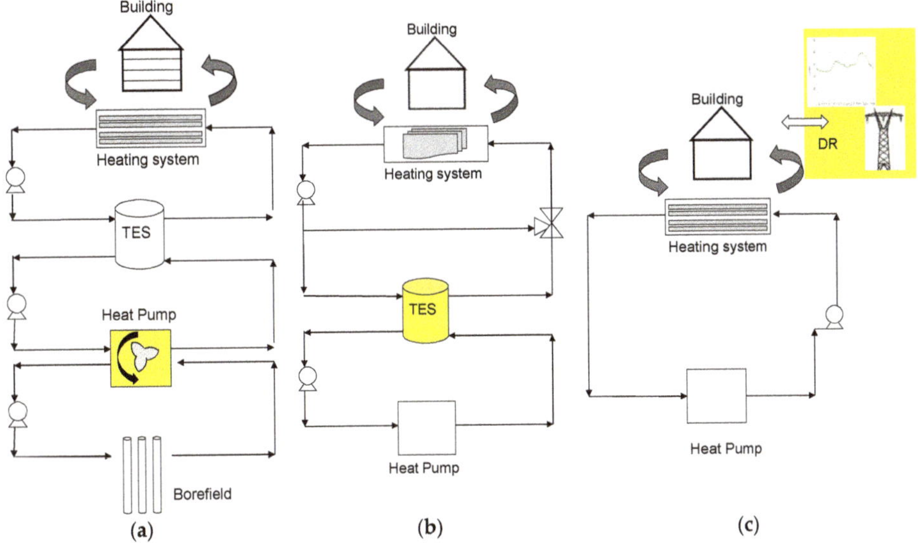

Figure 2. Schematics of the three case studies: (**a**) Case A: energy efficiency, (**b**) Case B: thermal storage, (**c**) Case C: demand response. TES = thermal energy storage; DR = demand response.

3.2.2. Case B: Energy Storage System

This case study was used to assess the energy flexibility potential of a storage system. A single dwelling of 100 m^2 with an air source heat pump (ASHP, represented by Type 941 in TRNSYS) and radiators (Type 1231) was considered (Figure 2b). Given the low thermal inertia of radiators, the energy storage system was necessary to guarantee a more stable operation of the HP without too many cycles. A thermally stratified water tank was used as an energy storage system (Type 4 has been implemented), which consisted of 20 fully-mixed equal volume segments where the model assumes an energy balance for each volume. The tank volume varied from a small buffer tank of 20 L up to 300 L. The building followed the existing Italian standards [63] for overall heat capacity of walls, roof, and windows, as reported in Table 2. Modern low temperature radiators, which can be supplied with hot water at temperatures varying between 40 $^\circ$C and 50 $^\circ$C on the basis of a compensation curve [33], were installed. The design heat pump capacity has been assessed at 3 kW, evaluated at the design outside temperature of $-1\ ^\circ$C and inside temperature of 20 $^\circ$C. The heat pump was controlled by the external ambient (<15 $^\circ$C), the tank temperature, and the radiators' surface temperature. Furthermore, an interval of at least 6 min between each switching operation, on or off, has to be guaranteed to avoid any excessively rapid cycling of the heat pump. The indoor temperature activates the pump supplying hot water to the radiators.

3.2.3. Case C: Demand Response

This case study considered a single dwelling of 100 m^2 with an air source heat pump and underfloor heating system (Figure 2c). The heat pump capacity has been assessed at 2 kW, evaluated as 70% of the design value at the design outside temperature of -1 °C and inside temperature of 20 °C, similar to the case in Section 3.2.1. Given the thermal inertia of the chosen emission system, a storage tank is not necessary in this configuration [33]. The heat pump was controlled by the external ambient temperature (<15 °C), the indoor air temperature, and the slab surface temperature, while a minimum time of 6 min was guaranteed between any on-off cycle. The analysis aimed at showing the influence of the price signal on the HP load profile and consequently on the energy bill for the final user. When the DR strategy is applied, the final user is subject to a variable price and the system is allowed to extend the comfort band of ±2 °C. On the basis of the electricity market price on a typical winter's day in Italy (2 January 2017 [64]), the temperature set-point was varied according to Figure 3: when the electricity market price was lower the temperature set-point was increased, when the electricity market price was higher the set-point was decreased, so to force pre-heating when the electricity is cheaper. The temperature set-point profile is a step curve which drives a rule based control (considering the thermal inertia of the considered system and the real implementation constraints for stable operation, a continuous variability of the set-point has been considered not useful).

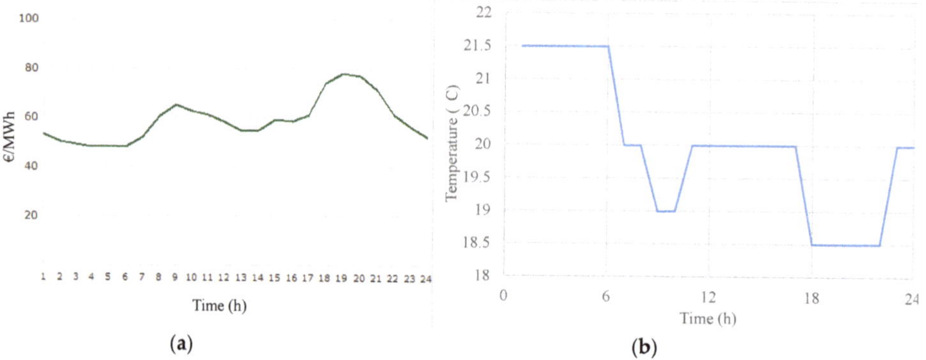

Figure 3. (**a**) Electricity market price on a winter's day (2 January 2017) and (**b**) daily internal temperature set-point profile.

4. Results and Discussion

4.1. Case A

In Figure 4, the heat pump electric power consumption in case of ON-OFF operation and variable capacity is shown. It is possible to see that in the ON-OFF configuration the HP power demand is higher and the device has to cycle several times. In the variable capacity configuration, the heating load can be adjusted on the basis of the real building heat demand that varies with the external ambient temperature. During the 24 h of operation represented in Figure 4a, indeed, the ambient temperature is always higher than the design value, then the HP modulates the thermal power provided. The variable capacity heat pump allows energy savings because the energy is produced following the real demand, then reducing the need to store it and the consequent thermal losses. For instance, in the month of January, an energy use reduction of about 15% is obtained for the considered case simulated in

TRNSYS. It is worth noticing that such energy saving is guaranteed even if in the variable capacity heat pump system, the auxiliaries (e.g., circulation pumps) run for longer periods.

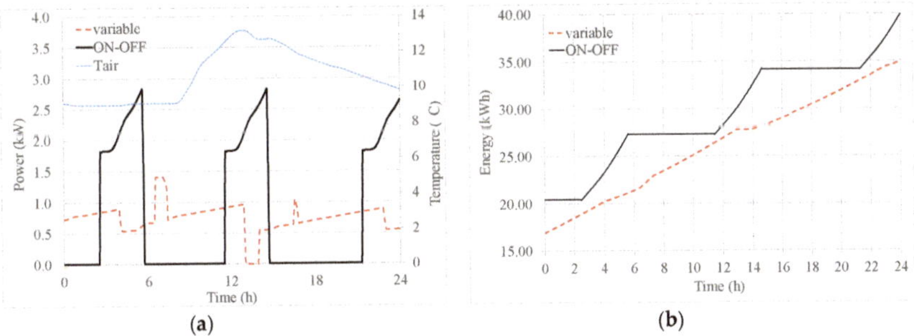

Figure 4. (**a**) Electric power profile and (**b**) cumulative energy curve of the heat pump (HP) in ON-OFF configuration and variable capacity configuration.

Figure 4b, instead, represents the cumulative energy use of the heat pump in the two configurations. In case of an ON-OFF regulation, the heat pump tends to concentrate the heating of the building in given amounts of time (i.e., pre-heating) and this confirms the higher energy thermal losses. The horizontal distance between the two curves of the cumulative energy represents the time shift of the energy use, which can also reach 8 h in the case represented in Figure 4b.

The load shifting indicator ($LS_{DSM,t}$) comparing the two configurations has been calculated. It is illustrated in Figure 5. As previously explained, the variable capacity HP stays on for a longer period of time with a lower power demand. This is shown also by the load shifting indicator: it is positive when the power is higher in the variable capacity configuration and negative when it is higher in the ON-OFF configuration. It has been calculated by means of Equation (2) and set at +100% when the variable capacity HP is on, while the ON-OFF HP is off. The electric power profile modification obtained with this energy efficient action is useful, for example, for peak shaving DSM strategies: the load can be reduced to 50% or even more. Moreover, in general, energy efficient solutions help energy conservation DSM strategies.

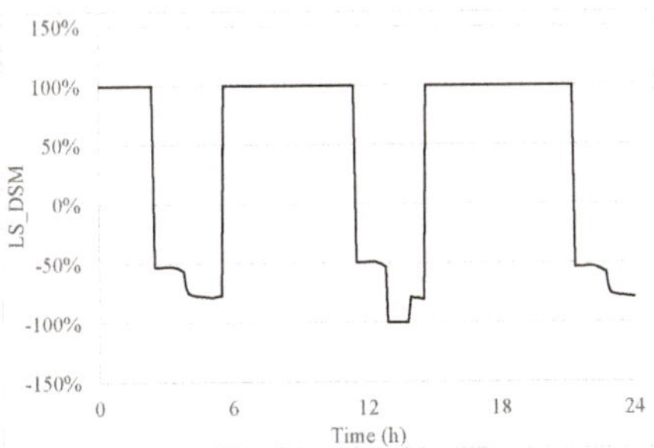

Figure 5. Load shifting indicator ($LS_{DSM,t}$) for a typical winter's day: comparison between the two configurations with (variable capacity HP) and without demand side management (DSM) (ON-OFF HP).

4.2. Case B

In Figure 6a the power consumption of the heat pump, by varying the TES size, is shown. The volume varies from a small buffer tank of 20 L up to 300 L. It is apparent that by decreasing the size of the water tank, the heat pump has to switch ON/OFF several times in order to satisfy the thermal energy demand of the building. A bigger tank indeed provides more energy autonomy, and then flexibility. Figure 6a highlights also that for proper operation of the heat pump system, a minimum tank volume of 100 L is preferable, then it is considered as a reference design configuration (i.e., without DSM). Figure 6b shows the cumulative energy curves of the heat pump in the reference configuration with a tank volume of 100 L and in the "DSM configuration" with a bigger storage tank of 300 L. It is evident that the system with a higher volume tends to pre-heat more and provides energy to the house for a longer period before the heat pump needs to be switched on again. Indeed, a 100 L and 300 L can provide, respectively, 1 kWh and 3 kWh of energy stored. The time shift between the two is about 2–4 h. Regarding the total energy use, there is not a consistent difference between the two selected tank volumes. However, the energy use tends to increase with the tank size because of more energy losses during the energy storage period.

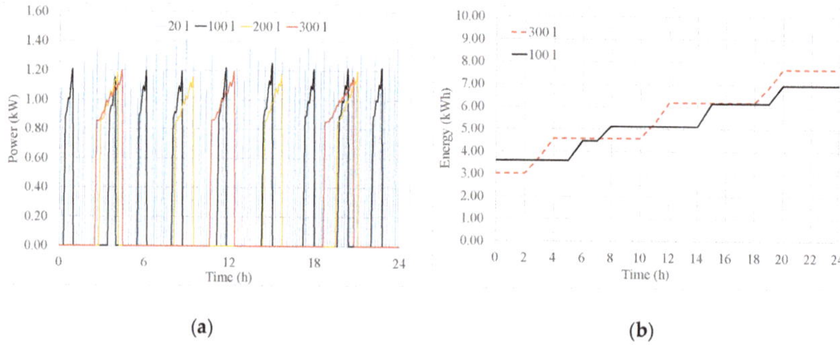

Figure 6. (**a**) Electric power profile and (**b**) cumulative energy curve of the HP for different tank sizes.

In Figure 7a the load shifting indicator is drawn when the tank size is increased from 100 L up to 300 L for 24 h of operation. Using a bigger tank, the heat pump switches on a few times for longer periods with a similar power demand in the two configurations. This solution is good for concentrating the heat pump working period in a given moment during the day, or for avoiding the operation in certain hours, for example, for night load shifting. In the latter case a bigger volume than those analyzed is needed. This is confirmed by Figure 7b which shows the load shifting indicator for a 1000 L tank compared to a 100 L tank. The HP can indeed be OFF for a long time period.

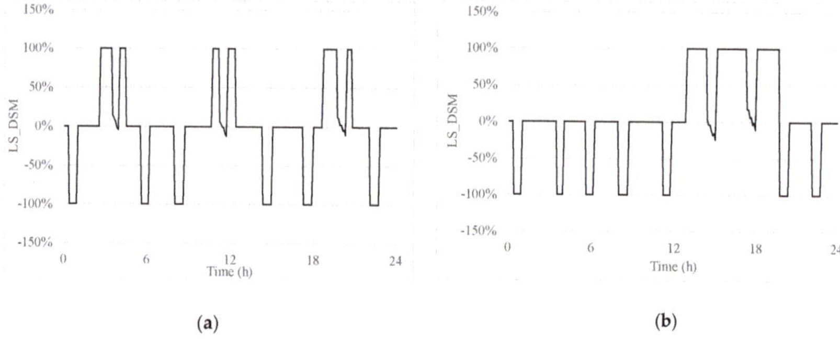

Figure 7. Load shifting indicator ($LS_{DSM,t}$).for the comparison between (**a**) a system with a 300 L vs. 100 L TES and (**b**) a system with 1000 L vs. 100 L TES.

4.3. Case C

In Figure 8a the power consumption trend of the heat pump is shown for the case without DR and for the case with DR activated. It is clear that when there is a variable temperature set-point, the heat pump switches on when the latter has higher values, that is, especially during the period 0–6 h. Instead, when the temperature set-point is decreased, the heat pump tends to work for shorter periods of time. The cumulative energy curve (Figure 8b) shows that during demand response programs, the heat pump pre-heats the house and for this reason more energy is used (about 2% for the month of January). The time shift allowed can reach up to 10 h and it represents energy use in advance rather than the reference case.

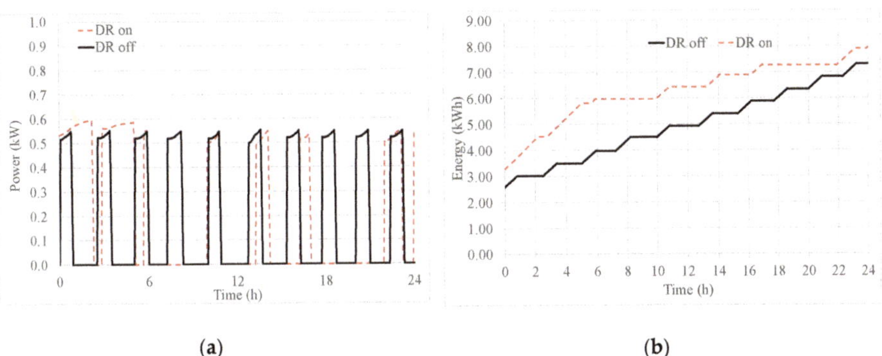

(a)   (b)

Figure 8. (**a**) Electric power profile and (**b**) cumulative energy curve of the HP with and without the DR program.

Figure 9 shows the load shifting indicator when the DR program is on or off during 24 h of operation on a winter's day. Comparing the load shifting trend with the internal temperature set-point, it is evident that the load is shifted to the hours when the set-point is higher, that is, the hours with cheaper electricity prices. Therefore, this DSM technology increases the load during off-peak hours, while decreasing it during on-peak hours.

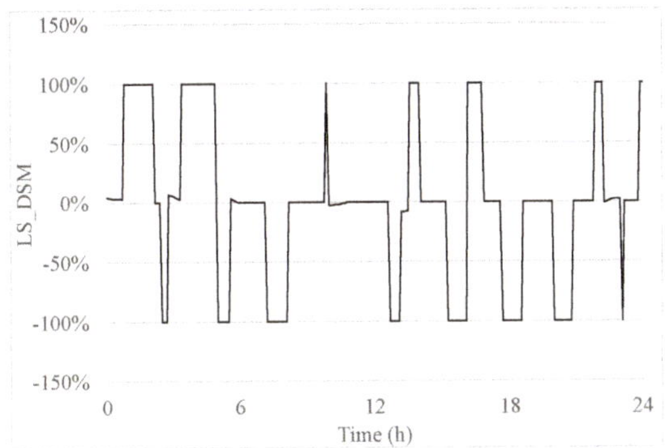

Figure 9. Load shifting indicator ($LS_{DSM,t}$) for the comparison between the system with or without the DR program.

4.4. Discussion

The presented case studies were intended to show the flexibility introduced by specific DSM technologies, namely:

1. energy efficient devices, represented by a more efficient variable capacity heat pump,
2. thermal storage, where different tank volumes are coupled with HP and radiators,

3. demand response, implemented by means of a variable internal temperature set-point which reflects a variable electricity price.

It is evident that each of these systems produces a modification of the electric power demand of the HP and provides flexibility. In more detail, looking at the HP power profile and at the LS_{DSM} indicator, the following conclusions can be drawn (see Figure 10):

1. Energy efficient DSM technologies allow energy conservation (reduction of the overall load) and, as a consequence, peak shaving (reduction of the maximum load). Indeed, the variable capacity HP is ON for longer periods but with a lower load.
2. Energy storage systems achieve load shifting, and the time shifting depends on the amount of energy that can be stored. Without any other external constraints, the storage volume affects the operation of the system and the time to which the load is shifted.
3. Eventually DR programs, based on real time pricing, use the price signal to induce the user to shift their load to hours with a lower electricity price. With DR programs, load shifting to off-peak hours (valley filling, i.e., building off-peak loads) and consequently peak shaving of the overall power system can be operated.

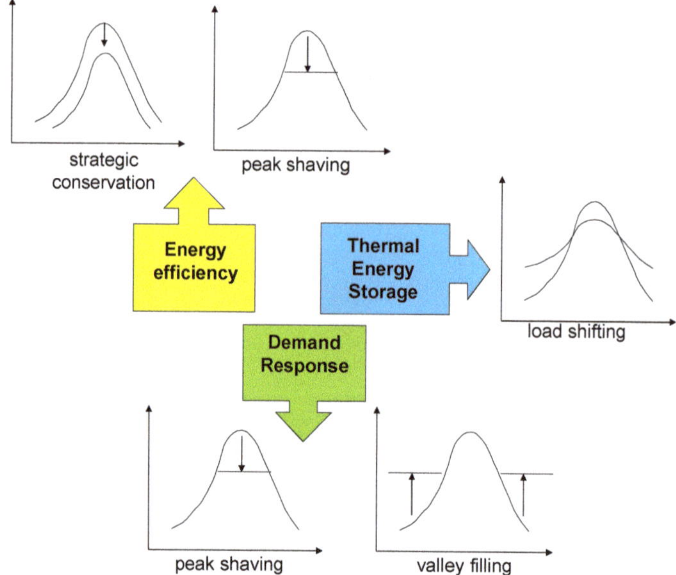

Figure 10. DSM technologies and the DSM strategies that they can help to implement.

The results also clarify the role of the thermal inertia available in the considered systems. The design and configuration of the analyzed cases have a huge impact on the final results. In particular, the thermal inertia provided by the building and emission system play a major role. When an underfloor heating system is used, and without any additional energy storage tanks, the time shifting in energy use is consistent. In Case A it can reach 8 h and represents a postponed use of energy, while in Case C it goes up to 10 h of energy used in advance, rather than the reference case. Looking at the load shifting indicator, in Case A the HP load is almost always present, even if with lower values (50% of the case without DSM technology or even lower). Whereas, by comparing Case B and C, in both cases the HP is ON less times for longer periods without reduction of the peak load. The difference between the two configurations is that in Case C, the periods when the HP works are actively controlled by the electricity price (the energy use occurs more during off-peak periods), while in Case B, unless in presence of external constraints, the load shifting depends only on the size of the tank and can occur at any time, depending on the operation of the system. For the configurations analyzed, for example, with the 300 L storage tank, the heat pump switches on three times in a day for 6 h in total, instead of four times for 4 h. Furthermore, even if the configuration with radiators and TES provides a shorter time shifting potential rather than the other systems, the relative power that can be actually shifted is big (always 100%). Eventually only Case A shows a potential energy use reduction, mainly due to the fact that when energy demand is shifted in time (i.e., Case B and Case C), it needs to be stored somehow thus increasing the energy losses.

In conclusion, the LS_{DSM} indicator provides knowledge about the dynamic behavior of a system where a considered DSM technology has been introduced. Its trends suggest the purpose of the DSM strategy (i.e., peak shaving, valley filling) and the time shifting indicator gives an idea about how much time is possible to deviate from the normal operation. The results discussed here are useful to understand the flexibility that can be expected by heat pumps in buildings and how it can be activated. These results are case specific and are related to the operation of the analyzed system: the flexibility available varies with time and is affected by what happened in the previous periods. Having similar indicators for a wide spread of technologies would allow quantifying the total flexibility that a given sector (e.g., TCL sector) can guarantee to the grid. However, in order to really understand how the flexibility can be exploited, it is necessary to model the dynamics of the overall power system with both the electricity demand side and production side. Only in this way is it possible to account for the interaction between the demand and the supply of electricity [65]. Such interaction is, for example, responsible for the payback load effect that could cause a new peak demand after a DR event or for the fact that the actual flexibility asked of every user is generally much lower than the total flexibility that the user could provide [47]. Indeed, the desynchronization and wide share of DSM events is paramount for an effective outcome for these technologies.

5. Conclusions

In this work, an overview of the demand side management strategies that can be implemented in the field of air conditioning and heat pumps is presented. The different DSM available technologies have been categorized into three groups: energy efficient devices, energy storage, and DR. In particular in this work, DSM strategies applied to heating systems using heat pumps are studied in detail in order to provide a qualitative quantification of the available flexibility and the expected behavior of such systems. Possible ways to assess the energy flexibility provided by heat pumps adhering to DSM strategies are discussed by means of three representative case studies. The flexibility is evaluated in terms of load shifting potential and time shifting of the energy use. The case studies consider the flexibility provided by: a heat pump efficient design (variable capacity HP), a thermal energy storage (thermally stratified water tank), and a DR program to drive the electricity consumptions towards off-peak hours. All the analyzed DSM technologies are effective, but they produce different effects on the basis of the system configuration. The main conclusions that can be drawn on the basis of the presented results are:

- Energy efficiency actions can produce mainly peak shaving and energy conservation. When such action is represented by a variable capacity heat pump, the load is always present but with lower values (50% or even less). In case of distribution systems with a high thermal inertia, such as underfloor heating, the energy use can be postponed for several hours (8 h in the analyzed case).
- Energy storage systems allow load shifting. The amount of energy that can be shifted depends on the storage size and, unless in presence of external constraints, the time to which the load is moved cannot be actively controlled. The time shifting in cases of low thermal inertia distribution systems, that is, radiators, is limited but the relative power that can be postponed is high (100% of the nominal value).

- DR operates an active load shifting to off-peak hours (valley filling) and peak shaving. In this case, energy is used in advance in comparison with the reference case. The time shifting is long in the presence of underfloor heating systems (for a 100 m^2 apartment it can be anticipated at 4 h). This kind of strategy tends to increase the overall energy consumption but allows a better operation of the power system avoiding high peak demand.

Author Contributions: Conceptualization and Writing-Original Draft Preparation, A.A.; Writing-Review & Editing, F.P.

Funding: This research was funded by MIUR of Italy in the framework of PRIN2015 project: "Clean heating and cooling technologies for an energy efficient smart grid", Prot. 2015M8S2PA.

Acknowledgments: The authors wish to thank the MIUR of Italy for supporting this study in the framework of PRIN2015 project: "Clean heating and cooling technologies for an energy efficient smart grid", Prot. 2015M8S2PA

Conflicts of Interest: The authors declare no conflict of interest. The funders had no role in the design of the study; in the collection, analyses, or interpretation of data; in the writing of the manuscript, or in the decision to publish the results.

References and Note

1. EIA (Energy Information Administration). How Much Energy is Consumed in Residential and Commercial Buildings in the United States? Available online: http://www.eia.gov/tools/faqs/faq.cfm?id=86&t=1 (accessed on 13 July 2018).
2. EU (European Union). Energy Efficiency-Buildings. Available online: https://ec.europa.eu/energy/en/topics/energy-efficiency/buildings (accessed on 13 July 2018).
3. Coulomb, D.; Dupont, J.L.; Pichard, A. *The Role of Refrigeration Industry in the Global Economy*; Report No. 29th Informatory Note on Refrigeration Technology; International Institute of Refrigeration: Paris, France, November, 2015.
4. Thomas, N.; Pascal, W. *European Heat Pump Market and Statistics Report 2015*; Technical Report for The European Heat Pump Association AISBL (EHPA): Brussels, Belgium, 2015.
5. Callaway, D.S.; Hiskens, I.A. Achieving controllability of electric loads. *Proc. IEEE* **2011**, *99*, 184–199. [CrossRef]
6. Jensen, S.Ø.; Marszal-Pomianowska, A.; Lollini, R.; Pasut, W.; Knotzer, A.; Engelmann, P.; Stafford, A.; Reynders, G. IEA EBC Annex 67 Energy Flexible Buildings. *Energy Build.* **2017**, *155*, 25–34. [CrossRef]
7. EU (European Union). Directive 2010/31/EU of the European Parliament and of the Council of 19 May 2010 on the Energy Performance of Buildings, 2010. Available online: https://eur-lex.europa.eu/legal-content/EN/TXT/PDF/?uri=CELEX:32010L0031&from=en (accessed on 13 July 2018).
8. Strbac, G. Demand side management: Benefits and challenges. *Energy Policy* **2008**, *36*, 4419–4426. [CrossRef]
9. Gellings, C.W. The smart grid. In *Enabling Energy Efficiency and Demand Response*; The Fairmont Press: Lilburn, GA, USA, 2009.
10. Harby, K. Hydrocarbons and their mixtures as alternatives to environmental unfriendly halogenated refrigerants: An updated overview. *Renew. Sustain. Energy Rev.* **2017**, *73*, 1247–1264. [CrossRef]
11. Bolajian, B.O.; Huan, Z. Ozone depletion and global warming: Case for the use of natural refrigerant —A review. *Renew. Sustain. Energy Rev.* **2013**, *18*, 49–54. [CrossRef]
12. Nawaz, K.; Shen, B.; Elatar, A.; Baxter, V.; Abdelaziz, O. R290 (propane) and R600a (isobutane) as natural refrigerants for residential heat pump water heaters. *Appl. Therm. Eng.* **2017**, *127*, 870–883. [CrossRef]
13. Zajacs, A.; Lalovs, A.; Borodinecs, A.; Bogdanovics, R. Small ammonia heat pumps for space and hot tap water heating. *Energy Procedia* **2017**, *122*, 74–79. [CrossRef]
14. Krishnamoorthy, S.; Modera, M.; Harrington, C. Efficiency optimization of a variable-capacity /variable-blower-speed residential heat-pump system with ductwork. *Energy Build.* **2017**, *150*, 294–306. [CrossRef]
15. Park, D.Y.; Yun, G.; Kim, K.S. Experimental evaluation and simulation of a variable refrigerant-flow (VRF) air-conditioning system with outdoor air processing unit. *Energy Build.* **2017**, *146*, 122–140. [CrossRef]
16. Aynur, T.N.; Hwang, Y.; Radermacher, R. Simulation comparison of VAV and VRF air conditioning systems in an existing building for the cooling season. *Energy Build.* **2009**, *41*, 1143–1150. [CrossRef]

17. Özahi, E.; Abusoglu, A.; Kutlar, A.I.; Dagcı, O. A comparative thermodynamic and economic analysis and assessment of a conventional HVAC and a VRF system in a social and cultural center building. *Energy Build.* **2017**, *140*, 196–209. [CrossRef]

18. Sumeru, K.; Nasution, H.; Ani, F.N. A Review on two-phase ejector as an expansion device in vapor compression refrigeration cycle. *Renew. Sustain. Energy Rev.* **2012**, *16*, 4927–4937. [CrossRef]

19. Wang, H.; Cai, W.; Wang, Y. Optimization of a hybrid ejector air conditioning system with PSOGA. *Appl. Therm. Eng.* **2017**, *112*, 1474–1486. [CrossRef]

20. Boccardi, G.; Botticella, F.; Lillo, G.; Mastrullo, R.; Mauro, A.W.; Trinchieri, R. Experimental investigation on the performance of a transcritical CO_2 heat pump with multiejector expansion system. *Int. J. Refrig.* **2017**, *82*, 389–400. [CrossRef]

21. Bansal, P.; Vineyard, E.; Abdelaziz, O. Status of not-in-kind refrigeration technologies for household space conditioning, water heating and food refrigeration. *Int. J. Sustain. Built Environ.* **2012**, *1*, 85–101. [CrossRef]

22. Kitanovski, A.; Egolf, P.W. Application of magnetic refrigeration and its assessment. *J. Magn. Magn. Mater.* **2009**, *321*, 777–781. [CrossRef]

23. Promoppatum, P.; Yao, S.-C.; Hultz, T.; Agee, D. Experimental and numerical investigation of the cross-flow PCM heat exchanger for the energy saving of building HVAC. *Energy Build.* **2017**, *138*, 468–478. [CrossRef]

24. Du, Z.; Jin, X.; Fang, X.; Fan, B. A dual-benchmark based energy analysis method to evaluate control strategies for building HVAC systems. *Appl. Energy* **2016**, *183*, 700–714. [CrossRef]

25. Ascione, F.; Bianco, N.; De Stasio, C.; Mauro, G.M.; Vanoli, G.P. A new comprehensive approach for cost-optimal building design integrated with the multi-objective model predictive control of HVAC systems. *Sustain. Cities Soc.* **2017**, *31*, 136–150. [CrossRef]

26. Kusiak, A.; Li, M.; Tang, F. Modeling and optimization of HVAC energy consumption. *Appl. Energy* **2010**, *87*, 3092–3102.

27. Kusiak, A.; Tang, F.; Xu, G. Multi-objective optimization of HVAC system with an evolutionary computation algorithm. *Energy* **2011**, *36*, 2440–2449. [CrossRef]

28. Garnier, A.; Eynard, J.; Caussanel, M.; Grieu, S. Predictive control of multizone heating, ventilation and air-conditioning systems in non-residential buildings. *Appl. Soft Comput.* **2015**, *37*, 847–862. [CrossRef]

29. Arteconi, A.; Hewitt, N.J.; Polonara, F. State of the art of thermal storage for demand side management. *Appl. Energy* **2012**, *93*, 371–389. [CrossRef]

30. Arteconi, A.; Polonara, F. Demand side management in refrigeration applications. *Int. J. Heat Technol.* **2017**, *35*, S58–S63. [CrossRef]

31. Dincer, I.; Rosen, M.A. *Thermal Energy Storage. Systems and Applications*, 2nd ed.; Wiley: Chichester, UK, 2011.

32. Hasnain, S.M. Review on sustainable thermal energy storage technologies, Part 2: Cool thermal storage. *Energy Convers. Manag.* **1998**, *39*, 1139–1153. [CrossRef]

33. Arteconi, A.; Hewitt, N.J.; Polonara, F. Domestic Demand-Side Management (DSM): Role of Heat Pumps and Thermal Energy Storage (TES) systems. *Appl. Therm. Eng.* **2013**, *51*, 155–165. [CrossRef]

34. Zhang, Y.; Zhou, G.; Lin, K.; Zhang, Q.; Di, H. Application of latent heat thermal energy storage in buildings: State-of-the-art and outlook. *Build. Environ.* **2007**, *42*, 2197–2209. [CrossRef]

35. Arteconi, A.; Costola, D.; Hoes, P.; Hensen, J.L.M. Analysis of control strategies for thermally activated building systems under demand side management mechanisms. *Energy Build.* **2014**, *80*, 384–393. [CrossRef]

36. Yin, R.; Kara, E.C.; Li, Y.; DeForest, N.; Wang, K.; Yong, T.; Stadler, M. Quantifying flexibility of commercial and residential loads for demand response using setpoint changes. *Appl. Energy* **2016**, *177*, 149–164. [CrossRef]

37. Fischer, D.; Madani, H. On heat pumps in smart grids: A review. *Renew. Sustain. Energy Rev.* **2017**, *70*, 342–357. [CrossRef]

38. Geidl, M.; Arnoux, B.; Plaisted, T.; Dufour, S. A fully operational virtual energy storage network providing flexibility for the power system. In Proceedings of the 12th IEA Heat Pump Conference, Rotterdam, The Netherlands, 15–18 May 2017.

39. Schwalbea, R.; Häuslerb, M.; Stiftera, M.; Esterl, T. Market driven vs. grid supporting heat pump operation in low voltage distribution grids with high heat pump penetration—An Austrian case study. In Proceedings of the 12th IEA Heat Pump Conference, Rotterdam, The Netherlands, 15–18 May 2017.

40. Bruninx, K. Improved Modeling of Unit Commitment Decisions under Uncertainty. Ph.D. Thesis, KU Leuven, Leuven, Belgium, 2016.

41. Rahmani-Andebili, M. Modeling nonlinear incentive-based and price-based demand response programs and implementing on real power markets. *Electr. Power Syst. Res.* **2016**, *132*, 115–124. [CrossRef]
42. Rahmani-andebili, M.; Shen, H. Energy Management of End Users Modeling their Reaction from a GENCO's Point of View. In Proceedings of the International Conference on Computing, Networking and Communications (ICNC), Silicon Valley, San Francisco, CA, USA, 26–29 January 2017.
43. Rahmani-andebili, M. Investigating Effect of Responsive Loads Models on Unit Commitment Collaborated with Demand Side Resources. *IET Gener. Transm. Distrib. J.* **2013**, *7*, 420–430. [CrossRef]
44. Baetens, R.; De Coninck, R.; Van Roy, J.; Verbruggen, B.; Driesen, J.; Helsen, L.; Saelens, D. Assessing electrical bottlenecks at feeder level for residential net zero-energy buildings by integrated system simulation. *Appl. Energy* **2012**, *96*, 74–83. [CrossRef]
45. Arteconi, A.; Ciarrocchi, E.; Pan, Q.; Carducci, F.; Comodi, G.; Polonara, F.; Wang, R. Thermal energy storage coupled with PV panels for demand side management of industrial building cooling loads. *Appl. Energy* **2016**, *185*, 1984–1993. [CrossRef]
46. Shen, B.; Ghatikar, G.; Lei, Z.; Li, J.; Wikler, G.; Martin, P. The role of regulatory reforms, market changes, and technology development to make demand response a viable resource in meeting energy challenges. *Appl. Energy* **2014**, *130*, 814–823. [CrossRef]
47. Arteconi, A.; Patteeuw, D.; Bruninx, K.; Delarue, E.; D'haeseleer, W.; Helsen, L. Active demand response with electric heating systems: Impact of market penetration. *Appl. Energy* **2016**, *177*, 636–648. [CrossRef]
48. Rahmani-andebili, M. Nonlinear demand response programs for residential customers with nonlinear behavioral models. *Energy Build.* **2016**, *119*, 352–362. [CrossRef]
49. Rahmani-andebili, M.; Shen, H. Price-Controlled Energy Management of Smart Homes for Maximizing Profit of a GENCO. *IEEE Trans. Syst. Man Cybern. Syst.* **2017**. [CrossRef]
50. Nuytten, T.; Claessens, B.; Paredis, K.; Van Bael, J.; Six, D. Flexibility of a combined heat and power system with thermal energy storage for district heating. *Appl. Energy* **2013**, *104*, 583–591. [CrossRef]
51. D'hulst, R.; Labeeuw, W.; Beusen, B.; Claessens, S.; Deconinck, G.; Vanthournout, K. Demand response flexibility and flexibility potential of residential smart appliances: Experiences from large pilot test in Belgium. *Appl. Energy* **2015**, *155*, 79–90. [CrossRef]
52. De Coninck, R.; Helsen, L. Quantification of flexibility in buildings by cost curves—Methodology and application. *Appl. Energy* **2016**, *162*, 653–665. [CrossRef]
53. Oldewurtel, F.; Sturzenegger, D.; Andersson, G.; Morari, M.; Smith, R.S. Towards a standardized building assessment for demand response. In Proceedings of the IEEE Conference on Decision and Control, Firenze, Italy, 10–13 December 2013; pp. 7083–7088.
54. Vanhoudt, D.; Geysen, D.; Claessens, B.; Leemans, F.; Jespers, L.; Van Bael, J. An actively controlled residential heat pump: Potential on peak shaving and maximization of self-consumption of renewable energy. *Renew. Energy* **2014**, *63*, 531–543. [CrossRef]
55. Reynders, G.; Diriken, J.; Saelens, D. Generic characterization method for energy flexibility: Applied to structural thermal storage in residential buildings. *Appl. Energy* **2017**, *198*, 192–202. [CrossRef]
56. Fischer, D.; Wolf, T.; Wapler, J.; Hollinger, R.; Madani, H. Model-based flexibility assessment of a residential heat pump pool. *Energy* **2017**, *118*, 853–864. [CrossRef]
57. Stinner, S.; Huchtemann, K.; Müller, D. Quantifying the operational flexibility of building energy systems with thermal energy storages. *Appl. Energy* **2016**, *181*, 140–154. [CrossRef]
58. Bianco, V.; Scarpa, F.; Tagliafico, L.A. Estimation of primary energy savings by using heat pumps for heating purposes in the residential sector. *App. Therm. Eng.* **2017**, *114*, 938–947. [CrossRef]
59. Select Software 7, Emerson Climate Technologies. Available online: http://www.emersonclimate.com/europe/en-eu/Resources/Software_Tools/Pages/Download_Full_Instructions.aspx (accessed on 13 July 2018).
60. Klein, S.A.; Beckman, W.A.; Mitchel, J.W.; Duffie, J.A.; Duffie, N.A.; Freeman, T.L. *TRNSYS Manual*; University of Wisconsin-Madison: Madison, WI, USA, 2009.
61. Salvala, G.; Pfafferott, J.; Jacob, D. Validation of a low-energy whole building simulation model. *Proc. SimBuild* **2010**, *4*, 32–39.

62. Ballarini, I.; Corgnati, S.P.; Corrado, V. Use of reference buildings to assess the energy saving potentials of the residential building stock: The experience of TABULA project. *Energy Policy* **2014**, *68*, 273–284. [CrossRef]
63. Italian Legislative Decree 311/2006. Disposizioni Correttive ed Integrative al Decreto Legislativo 19 Agosto 2005, n. 192, Recante Attuazione Della Direttiva 2002/91/CE, Relativa al Rendimento Energetico Nell'edilizia; Gazzetta Ufficiale Serie Generale n.26 del 01-02-2007 - Suppl. Ordinario n. 26, Roma, Italy, 2007.
64. Gestore del Mercato Elettrico. Available online: http://www.mercatoelettrico.org/it/ (accessed on 13 July 2018).
65. Patteeuw, D.; Bruninx, K.; Arteconi, A.; Delarue, E.; D'haeseleer, W.; Helsen, L. Integrated modeling of active demand response with electric heating systems coupled to thermal energy storage systems. *Appl. Energy* **2015**, *151*, 306–319. [CrossRef]

Article

Potential of District Cooling Systems: A Case Study on Recovering Cold Energy from Liquefied Natural Gas Vaporization

Alice Mugnini [1], Gianluca Coccia [1], Fabio Polonara [1,2] and Alessia Arteconi [1,*

[1] Dipartimento di Ingegneria Industriale e Scienze Matematiche, Università Politecnica delle Marche, Via Brecce Bianche 1, 60131 Ancona, Italy

[2] Consiglio Nazionale delle Ricerche, Istituto per le Tecnologie della Costruzione, Viale Lombardia 49, 20098 San Giuliano Milanese (MI), Italy

* Correspondence: a.arteconi@univpm.it; Tel.: +39-071-2204760

Received: 30 June 2019; Accepted: 3 August 2019; Published: 6 August 2019

Abstract: District cooling systems (DCSs) are networks able to distribute thermal energy, usually as chilled water, from a central source to industrial, commercial, and residential consumers, to be used for space cooling/dehumidification. As cooling demand will increase significantly in the next decades, DCSs can be seen as efficient solutions to improve sustainability. Although DCSs are considered so relevant for new city developments, there are still many technical, economic, and social issues to be overcome to let such systems to spread out. Thus, this paper aims to highlight the advantages and issues linked to the adoption of DCSs for building cooling when cold is recovered from a specific application. A case study based on liquified natural gas (LNG) cold energy recovery from the transport sector is presented. Starting from the estimation of the free cooling availability, a DCS design method is proposed and the potential energy saving is investigated. Results show that a DCS using the cold waste derived from LNG can provide a relevant amount of electricity saving (about 60%) for space cooling compared to traditional solutions, in which standard air conditioning systems are installed in every building.

Keywords: district cooling; liquid to compressed natural gas; thermal energy storage; energy efficiency; LNG

---

1. Introduction

Buildings are responsible for about 40% of worldwide energy use [1], and their cooling demand is increasing. Global residential space cooling demand was estimated to be about 300 TWh in 2000, and it is expected to rise up to 4000 TWh in 2050 and 10,000 TWh in 2100 [2]. According to different prognoses, in European Union (EU) countries the cooling demand will increase of almost 60% in 2030 respect to nowadays [3,4]. These statistics are even more valid in regions with tropical climates [5]. For instance, the percentage of energy used for space cooling in Mumbai (40%) is nearly as twice as that of London (20%) [6].

Providing cooling in an efficient way is harder than providing heat because the cooling demand is more difficult to predict than the heat demand, particularly the peaks [4]. Cooling demand, in fact, depends on different factors that can change quickly, such as solar radiation, internal heat gains, and the urban heat island effect [3]. Moreover, due to the increasing renewable energy source (RES) share, the stability of the electricity grid is at risk [7]. RESs such as wind and solar energy are intermittent, therefore they cannot be easily controlled and predicted. For this reason, it is harder and harder to keep the balance between generation and use of electricity, with an associated increasing risk of black-outs and other technical issues. Given all the aforementioned reasons, it is fundamental developing more efficient cooling systems to improve the sustainability of future cities.

Cooling demand can be met using individual solutions (air-conditioning split systems, or more efficient solutions such as absorption cycles and electric chillers), or using one solution for the whole building or cluster of buildings [5]. In this last case, district cooling systems (DCSs) represent an example of efficient means to fulfill the cooling demand [5]. A DCS can be considered as a centralized method of cold thermal energy production and distribution for space and process cooling. A typical DCS includes the following elements: a central generation unit, a distribution network, customers and a heat rejection system. Given its cost-effectiveness, chilled water is often used as heat transfer fluid.

The EU Energy Efficiency Directive [8] has stated that DCSs are one of the important pillars for achieving the energy efficiency target of reducing primary energy consumption by 20%, while the Strategic Energy Technologies Information System (SETIS) [9] has recognized DCS as Best Available Technology (BAT) for the cooling market in the EU. Moreover, the Heat Roadmap Europe [10,11] considers thermal networks (which include DCSs) to be fundamental to increase energy efficiency. Current yearly district cold deliveries can be estimated to be around 300 PJ worldwide (200 PJ in the Middle East, 80 PJ in the USA, 14 PJ in Japan, 10 PJ in Europe) and, in particular in Europe, there are around 150 systems in operation [12].

DCSs are becoming increasingly popular in the areas with high density of buildings, because of their high efficiency and flexibility [13]. However, literature on DCSs is limited because district cooling development has been more recent respect to district heating. Moreover, it has to be considered that the temperature difference between the supply water temperature and the return water temperature for DCSs is usually around 8 °C, while in district heating systems the difference is generally higher than 40 °C [3]. Additionally, the tendency to be cautious in the design phase leads often to oversize DCSs, thus resulting in equipment underutilization, which is detrimental for energy efficiency [6]. For all these reasons, DCSs show an increase of the piping system cost and of the required pumping energy respect to district heating systems [3]. Therefore, the amount of cold supplied by DCSs in the world is much smaller than heat supplied by district heating systems [12].

DCSs can be successfully integrated with RESs, trigeneration systems and thermal storage systems (TESs). Renewable energy resources include solar energy, wind energy, geothermal energy, biomasses, energy from the surface water (sea, rivers, lakes), etc. The integration of DCSs with renewable energy technologies allows reducing greenhouse gas emissions greatly [13]. DCSs using solar energy can be exploited in absorption refrigerating cycles [14]. Systems coupling DCSs and geothermal sources are mainly connected to aquifers or groundwater [13,15,16], while biomasses are generally based on municipal solid waste whose incineration and consequent heat is recovered for absorption chillers [13,17]. Surface water systems are frequently used in DCSs, using as cold source seawater heat pumps [18,19], river water [20], and deep lake water [21].

DCSs integrated with trigeneration systems, instead, have the possibility of improving the fuel energy efficiency, and are generally coupled with district heating systems to meet the heating and cooling demand of the users. Often, the DCS serves as a supplemental system to meet the cooling demand of the area [22], and thermal driven chillers are generally used to recover low grade heat from the trigeneration system [13]. Applications of single DCSs coupled with trigeneration systems are rarely reported, probably because the trigeneration energy efficiency is not high when the primary aim is to meet the cooling demand [13].

When integrated with TESs, DCSs energy consumption and operation cost at peak hours can be considerably reduced. TESs are able to store cold energy during periods of low cooling demand, and the stored energy can be used later to meet the required cooling load. In this way, the utilities benefit from the reduction of the peak electricity generation and consumers can take advantage of lower billing charges [13]. Typical TESs used in DCSs include chilled water, ice and phase change materials. Chilled water is generally used in its sensible form and its advantages result in low cost and high thermal capacity. Ice, instead, is used in form of latent heat, and is generally chosen when smaller storage volumes are required. Phase change materials (PCMs) adopted in DCSs usually include inorganic salt hydrates and paraffins. Salt hydrates have high latent heat during solidification/melting and are low-cost, but they can show incongruent melting and corrosion issues. Paraffins, on the other hand, can result in very high cost [13].

Another possible source rather than RES for DCSs is waste cold energy, which can be recovered from industry, commercial, and transport sectors [23]. While solar, geothermal and biomass waste energy are not as used as in district heating systems because of poor efficiencies due to energy conversion [13], liquefied natural gas (LNG) is in this study considered as a suitable source for DCSs. LNG is widely used to transport natural gas, especially when the distance between the production site and the market is longer than 2000 km [24]. LNG is obtained by cooling natural gas to −162 °C at atmospheric pressure; one cubic meter of LNG contains around 600 cubic meters of natural gas, making the energy density of LNG significantly higher [25]. Before being sent to the customers, LNG is re-gasified and a large quantity of cold energy can be released. A typical liquefaction process requires around 2900 kJ kg^{-1} of energy; of this amount, 2070 kJ kg^{-1} are dissipated as heat, while 830 kJ kg^{-1} are stored in LNG as cold [26]. LNG can also be used as vehicle fuel in its liquid form or re-gasified to be used as compressed natural gas (CNG). In the latter case, the liquid to compressed natural gas (L–CNG) refueling station has higher efficiency than traditional CNG refueling stations [27].

In these peculiar fuel stations, natural gas is stored in its liquid form and the CNG to refuel vehicles is obtained from LNG vaporization. The LNG is typically stored in a low-pressure cryogenic tank and then pumped with a cryogenic pump to increase the pressure of LNG up to the CNG pressure. Then, a high-pressure atmospheric vaporizer converts LNG into CNG, stored in pressurized cylinders. L–CNG refueling plants present a number of advantages: (i) the possibility to distribute CNG when no grid is available nearby; (ii) a higher fuel purity, since LNG is already purified at the liquefaction stage; (iii) a reduced operational cost compared to a standard compressor solution, thanks to the power saving from the LNG pumping; (iv) the possibility to supply also directly LNG to vehicles using it as fuel [28].

Nowadays, in Italy there are 53 LNG and CNG refueling stations powered by LNG [29]. Among these, 13 are L–CNG plants [30]. In these plants, the LNG vaporizer is typically a high-pressure atmospheric heat exchanger. Looking to the data on the CNG annual consumption for the automotive sector, a great waste of cold energy can be estimated when LNG is vaporized. Data available for 2015 reports an Italian LNG demand from L–CNG refueling stations of about 5400 t [31], thus about 1245 MWh year^{-1} of cold energy could be recovered from such plants.

Based on the knowledge of the authors, in literature there is no study of DSCs using waste cold energy derived from L–CNG refueling stations. Thus, in order to limit environment warming and to maximize energy efficiency, this work aims to evaluate the potential of such cold energy recovery to meet a residential district cooling demand. Since the cooling potential of vaporized L–CNG represents a free cooling source, its efficient exploitation has the possibility to greatly reduce the electric demand of the air-conditioning sector. Furthermore, the combined presence in the urban environment of residential and transport sectors highlights the potential benefits of their optimized integration.

The manuscript is structured as follows: Section 2 presents the case study considered and the description of the DCS design method. Section 3 describes the results of the study and tries to highlight the potential of the coupling in terms of electricity saving, taking as a baseline the same residential neighborhood satisfied with independent cooling systems. Finally, the main conclusions of the work are reported in Section 4.

2. Case Study

In this section, an application case is presented in which the cooling energy requirement of a residential neighborhood is satisfied with a DCS that uses the cold energy coming from a L–CNG refueling plant vaporizer (Figure 1). The following subsections provide the description of the cooling supply side and of the DC plant.

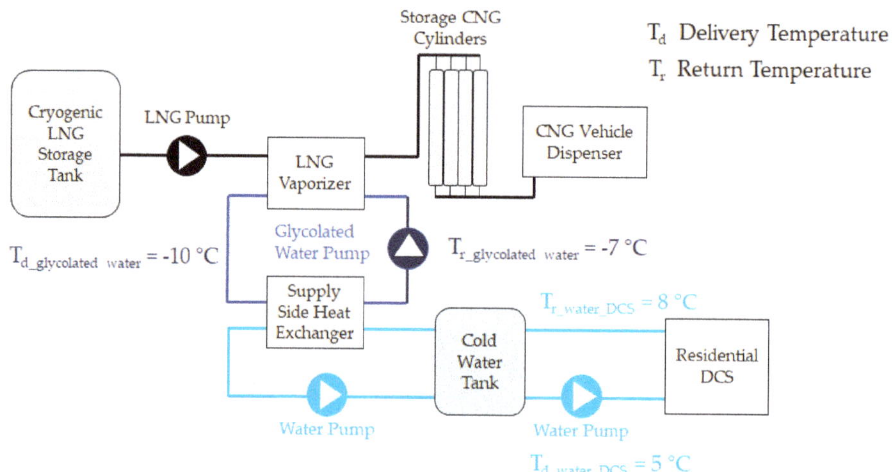

Figure 1. Plant scheme: cooling energy recovery from liquid to compressed natural gas (L–CNG) vaporizer to fulfil a residential district cooling systems (DCS) (focus on the cooling supply side).

2.1. L–CNG Cold Supply Side

To estimate the potential daily availability of recoverable cold energy from a L–CNG plant in a refueling station, data of CNG end-uses in Italy were considered. Franci [31] reports that the LNG consumption for the automotive sector in L–CNG plants is about 5400 t year^{-1}. Given such annual LNG consumption and assuming a quantity of energy of about 830 kJ kg^{-1} of LNG wasted during the LNG regasification process [26], about 380 kWh$_t$ day^{-1} of cold energy can be recovered if the fuel station is in operation every day of the year (i.e., 365 days per year); excluding the weekends the amount could result in about 550 kWh$_t$ day^{-1} (i.e., 253 days per year).

To evaluate the DCS capability to adapt its demand to the supply, the available cold energy is provided to the DCS by means of three different daily cooling power profiles (Figure 2). In the first case (Figure 2a, P1), the cooling power is constant and available 10 h a day, for every day of the year. This situation describes the typical opening and closing times of the Italian fuel stations, which were assumed to be from 7 a.m. to 7 p.m., and represents the simplest condition for the DC plant. Considering that in Italy many CNG fuel stations are closed during lunch time and weekends, another cooling power profile was considered accordingly in Figure 2b (P2). The last case, shown in Figure 2c (P3), introduces a variable profile with a peak power in the evening. In fact, as Farzaneh-Gord et al. [32] highlighted in their analysis by monitoring the number of refueling CNG vehicles in a day, a peak is generally observed between 6 and 7 p.m. This last scenario is therefore introduced to consider a variable cooling power during the day.

In the three profiles, the cooling power values were calculated by dividing the daily cold energy availability for the time during which the refueling station is supposed to be open. In the last case (P3), the peak size was chosen to satisfy the peak of the users' demand (about 100 kW$_t$, as explained in Section 2.2.2), while the cooling power in the other hours was determined by a constant distribution of the remaining daily cold energy.

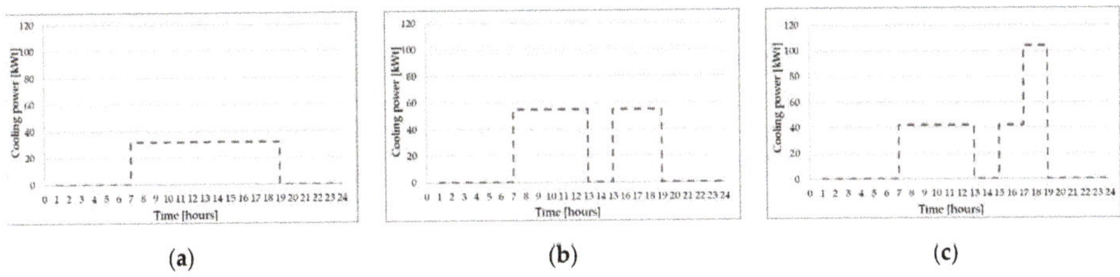

Figure 2. Daily cooling power recovery from a L–CNG fuel station. (**a**): constant profile for a station working every day of the year (P1); (**b**): constant profile for a station not working during weekends and stopping during lunch time (P2); (**c**): variable profile for a station not working during weekends and stopping during lunch time (P3).

The cooling power is supplied to the DCS by means of a heat exchanger ("supply side heat exchanger" in Figure 1). The heat transfer between the LNG and the water glycol was not directly modeled, but the free cooling is considered carried by a heat transfer fluid (water glycol mixture) supposed available at a temperature of −10 °C. The corresponding mass flow rate of the mixture was determined assuming an average temperature difference of 3 °C.

2.2. Residential District Model

A hypothetical residential neighborhood, located in Rome, Italy (41°55′ N, 12°31′ E), was modeled in TRNSYS [33]. To include the variability of external climatic conditions, a Meteonorm weather file was used. The number of users to be connected to the DCS has been determined through an estimation of the users' daily cooling energy requirements. Then, the minimum number of buildings whose base cooling demand is covered by the daily availability of free cooling was selected. This choice was made to ensure the maximum utilization of the free cooling.

The following Section 2.2.1 reports the details of the single user, while in Section 2.2.2 the DC plant sizing is explained.

2.2.1. Single User

The single user was modeled with Type 88 of TRNSYS. It is composed of a single thermal zone with a simple lumped capacitance structure. The main features of the envelope were extrapolated from TABULA Project [34] and a single-family house (SFH) building with the most recent construction year class (buildings built from 2006 onwards) was chosen as reference building. In the simulation model, however, only global data are required as input; thus, an overall building loss coefficient, which considers the different thermal characteristics and dimensions of opaque and transparent envelope components, was calculated (0.38 W m^{-2} K^{-1}). The single building is composed of two floors and has a living area of 160 m^2 (80 m^2 per floor). Based on the data reported in [34], for each external wall, 8% of window surface area rate was considered.

The modeled thermal gains are: (i) solar radiation, (ii) artificial lighting and (iii) occupants' gains. Solar radiation was provided as internal gain to Type 88 and was modeled starting from the total tilted surface radiation data for each building surface orientation. Then, the thermal power due to solar radiation actually entering the building was calculated and implemented according to the procedure reported in the Italian standard UNI TS 11300:1 [35]. Artificial lighting accounts for a power density of 5 W m^{-2} and it was assumed reasonably that it turns on if the total horizontal radiation is less than 120 W m^{-2}, while it turns off if radiation is greater than 200 W m^{-2}. As concerns the occupants' contribution, an occupancy density of 40 m^2 per person with a corresponding internal gain of 110 W per person was considered [36]. The contribution of natural ventilation was introduced, too, assuming air changes per hour equal to 0.5 h^{-1} [35].

The design peak cooling demand was evaluated by means of the Carrier-Pizzetti technical dynamic method [37,38]. This method allows estimating the summer thermal load (sensible and latent contribution) at different hours of the day, having as inputs the geographical coordinates, the envelope properties (mass and thermal transmittance) and the maximum external temperature: the maximum value was assumed as peak cooling demand. For Rome, where the outside design temperature suggested by [39] is 34 °C with a daily temperature variation of 11 °C, a sensible thermal load of 6.3 kW$_t$ was calculated (an internal temperature and relative humidity of 26 °C and 50%, respectively, were used for the assessment). The latent contribution is equal to about 2 kW$_t$.

The single user cooling distribution system comprises fan coil units (FCUs) modeled with Type 996, and water was considered as heat transfer fluid. Cold water is supplied at a temperature of about 10 °C, with a design temperature difference between delivery and return of 5 °C [40]. The control strategy aims to maintain the internal temperature set-point 24 h per day.

2.2.2. DCS Plant Sizing

From the energy simulation of the single user cooling demand for the main summer months (July and August [39]), a base cooling demand of about 40 kWh$_t$ was obtained. Therefore, to ensure that the cooling energy released by the L–CNG vaporizer can always satisfy the district base demand (Figure 3), it was calculated that 14 users (with the same demand) have to be connected to the DCS.

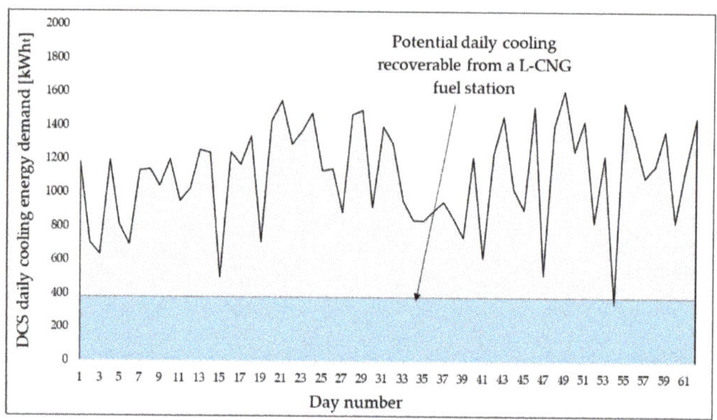

Figure 3. District daily cooling energy demand (July and August).

In order to cover higher energy demands or to cool buildings on days when there is no free cooling availability, every user was equipped with a backup system. The system includes a reversible air-to-water heat pump connected to the fan coils water circuit. The single heat pump is activated in cooling mode when the indoor temperature set-point goes above 26 °C: it integrates the energy removal when the DCS cooling power is not enough, and it meets all the cooling when no free-cooling is available.

Since there may be days in which there is no cold availability from the network (e.g., weekends), the heat pumps were sized to cover the entire design sensible cooling load (6.3 kW$_t$ for each user, 88.2 kW$_t$ for the overall neighborhood). Type 941 was used to model the backup system and its performance was derived from a manufacturer catalogue [41]. Figure 4 depicts, in a schematic way, how the backup system is connected to the user distribution system. The figure also shows the connection between the water fan coil units circuit and the DCS. The connection is realized with a heat exchanger modeled as a simple constant effectiveness device (Type 91 in TRNSYS, which adopts the ε-NTU method). To uncouple the cooling demand from the supply, a sensible thermal energy storage (TES) was also introduced and modeled through Type 4 as a stratified cold-water tank. Different sizes of the tank were tested, in order to find the storage that guarantees the required cold energy at the design cooling load for different periods of time.

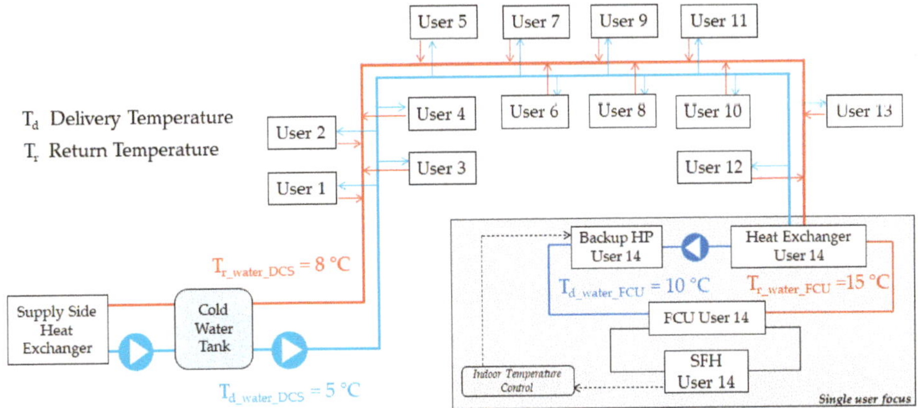

Figure 4. Residential district cooling network scheme (see Figure 1 for the details of the supply side). Focus on a single user configuration.

In the DCS, water was used as heat transfer fluid. It is delivered from the tank at a temperature of 5 °C, with a design temperature difference between supply and return of 3 °C. This choice was made to allow an efficient heat transfer between the working fluids at the single user heat exchanger, where FCUs are designed to be supplied at 10 °C (see Section 2.2.1).

The pipes were modeled as underground types (ground temperature of 14 °C) with Type 31. Referring to real pipes used for district heating and cooling systems [42], larger thicknesses (26–28 cm) of thermal insulation based on polyurethane rigid foam (thermal conductivity of 0.027 W m^{-1} K^{-1}) were considered to minimize the ground cooling energy losses. The sizing of the various DCS sections was realized maintaining the water circulation speed between 1.5 and 2 m s^{-1} in the main sections, and around 1 m s^{-1} in the secondary branches [40].

3. Results and Discussion

In this section, the simulation results of the coupling between the DCS and the L–CNG vaporizer are discussed. A reference case is introduced to evaluate the energy performance of the application. The reference case is represented by the same residential district in case of separate cooling systems: fan coil units powered by the same heat pump used for the backup cooling. From the energy simulation for the summer warmer months (July and August [39]), a total cold energy requirement of about 47 MWh$_t$ and an electrical consumption of about 17 MWh$_e$ were determined.

The first analysis evaluates the capability of the DCS to adapt its cooling demand to the free cooling power availability. A preliminary case, in which the DCS is directly connected to the supply side heat exchanger (without the thermal energy storage), was evaluated and the three profiles of free cooling power reported in Figure 2 were applied to the supply side.

In general, results show that the case study has a great energy recovery potential. Figure 5 represents how the users' cooling energy requirements are satisfied during the warmer summer months in case of different free cooling power profiles. In all cases the coverage of the buildings cooling demand exceeds 49%. However, although in a slight way, the shape of the free cooling profile seems to influence the district energy behavior. If the DCS is supplied with a constant cooling power profile (P1) every day, 50% of the user cooling demand is covered by the DCS itself and the electricity consumption is reduced by 52% compared to the reference case. With the increasing of the free cooling power capability (profile P2) during weekdays, a better exploitation of the cold energy provided by the DCS is obtained and the electricity saving becomes 58%. Introducing a most variable cooling power profile (P3) this improvement seems to be mitigated. Due to the daily mismatch between the cooling demand and supply, the P3 higher cooling peak power cannot be well exploited. Figure 6 shows the typical daily cooling demand of the district in the reference case. The thermal power demand starts around 10.00 a.m., when the P3 power availability is indeed low, and stays constant for several hours.

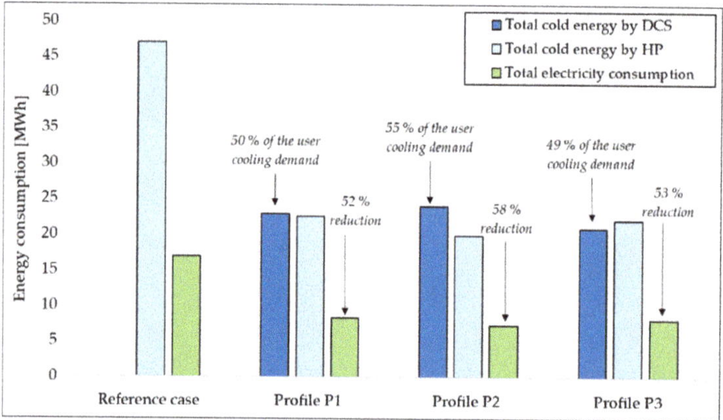

Figure 5. Total district electric consumption and users' cold energy demand divided into the share provided by DCS and by the backup systems (i.e., HP) with different free cooling power profile (months of July and August).

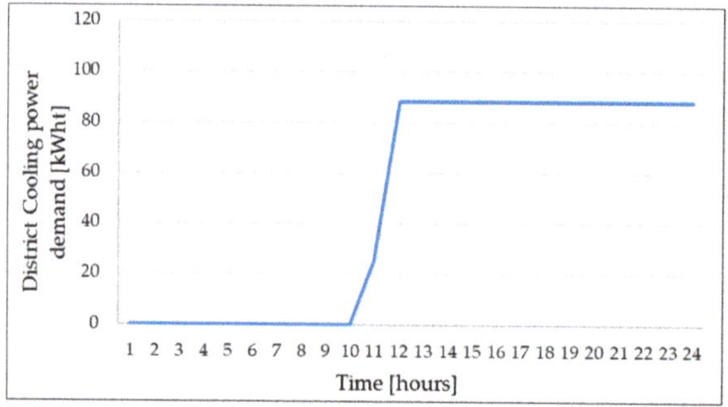

Figure 6. Total daily district cooling power demand in the reference case.

Therefore, it is evident that to minimize the electricity used for backup systems, a sufficiently high and continuous free cooling power is beneficial (i.e., P2). The effect of the introduction of a certain level of thermal inertia at the demand side is investigated in order to evaluate if it helps to decouple the cooling demand and the supply both in terms of time and of peak power potential. The installation of a tank of 60 m³, which in design conditions can provide the users' peak power demand (about 88 kW) for 1 hour with a temperature difference of 3 °C, is considered. Figure 7 shows how the users' cooling energy requirements are satisfied during the warmer summer months in case of different free cooling power profiles when the TES is introduced at the demand side. Comparing them to the data shown in Figure 5, a good increase in the DCS energy performance can be noticed in case of P3 profile: the seasonal demand covered by the DCS goes up to 59% with an electricity consumption reduction of about 60% compared to the reference case. For the other two cases (P1 and P2), the storage has practically no contribution, being their trend much more regular.

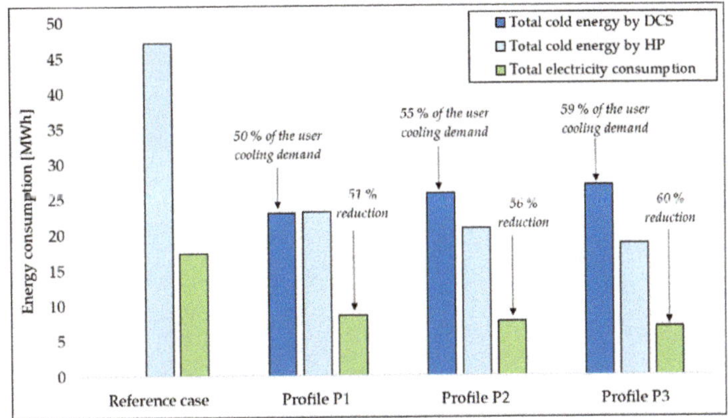

Figure 7. Total district electric consumption and users' cold energy demand divided into the share provided by DCS and by the backup systems (i.e., HP) with different free cooling power profile (months of July and August) in presence of a cold-water tank of 60 m^3.

The addition of a TES allows a better daily balance of energy demand and supply, too. Figure 8 shows for a typical summer day how the total cooling demand is satisfied when the free cooling availability is directly connected to the demand (Figure 8a) or decupled with a 60 m^3 TES (Figure 8b). Without the TES, the users have to switch on the backup systems from 5 p.m. onwards to maintain the temperature set-point, while in presence of the TES the backup systems do not switch on because more cooling power is available thanks to the stored energy.

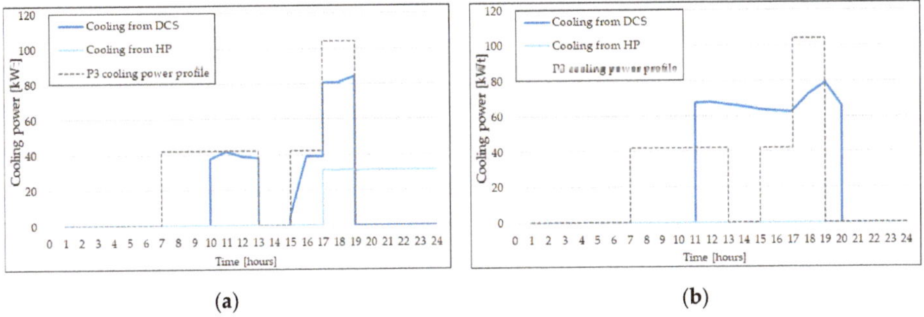

(a)　　　　　(b)

Figure 8. Daily cooling demand breakdown into the DCS and HP share in case of cooling supply profile P3: (**a**) without a TES; (**b**) in presence of a cold-water tank of 60 m^3.

Even though in the analyzed case the use of a thermal storage allows a higher level of energy recovery with higher electricity saving, in the DC plant sizing special attention must be paid to the choice of the tank volume. Tanks too small or too large, in fact, could cause performance degradation. Figure 9 reports the cold energy recovery from the L–CNG vaporizer compared to the global cooling demand and the percentage of electricity saving for different volumes of the TES (P3 profile). The best energy performance was obtained for tanks with a size between 30 and 90 m^3. Lower values do not provide enough level of thermal inertia, while higher values seem to slow down the dynamics of the system too much. In fact, if the tank is empty and the cooling power demand is at its peak, the large thermal inertia introduced by the tank prevents from satisfying the demand because the TES has to be first recharged; therefore, backup systems need to be switched on. Moreover, increasing the size of the tank enlarges the heat losses towards the environment, thus resulting in an energy efficiency penalization of the system.

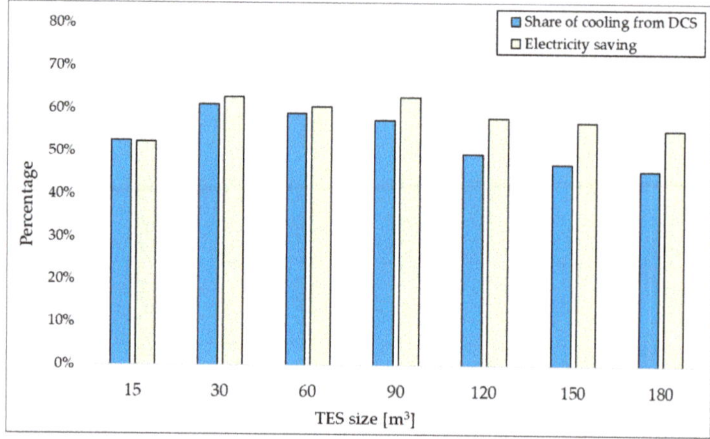

Figure 9. Percentage share of cooling from DCS to meet the total users' demand and electricity saving for different thermal energy storage (TES) sizes.

4. Conclusions

The present study provided an overview of advantages, issues and potential applications of district cooling systems (DCSs). DCSs can deliver cold energy supply at low cost and with low carbon dioxide emissions, and can be integrated with renewable energy systems, trigeneration systems and thermal energy storages. They can also recover efficiently waste cold energy from applications such as liquid to compressed natural gas (L–CNG) refueling, technology that was considered as a case study in this work.

Specifically, an energy system comprising a L–CNG refueling station vaporizing LNG to meet the space cooling demand of a small residential neighborhood was designed and modeled in a dynamic simulation environment. Results show that the DCS is able to provide high electricity saving levels in comparison with the baseline case, in which standard cooling systems are installed in every building. Regardless of the supply cooling power profile, electricity consumption reductions greater than 50% were obtained, reaching 60% in the best scenario (profile P3 with a 60 m^3 thermal energy storage at the demand side). It is necessary to highlight that these percentages are strongly dependent on the assumptions made to design the DCS, having assumed the cold energy availability from the LNG vaporizer as the district energy base demand. Further investigations should be conducted to evaluate the impact of the DCS sizing method used to select the number of users to be connected and to maximize the exploitation of the free cooling source.

To conclude, this work highlights that cold energy recovery from liquified natural gas vaporization in refueling stations can be effectively used to supply a residential DCS. Furthermore, since both the transport and the residential sector are included into the urban environment, the potential benefits of this integrated system are evident.

Author Contributions: Conceptualization, A.A.; methodology, A.M. and G.C.; validation, A.M.; writing—original draft preparation, A.M. and G.C.; writing—review and editing, G.C.; supervision, A.A. and F.P.; funding acquisition, F.P.

Funding: This research was funded by the Italian Ministero dell'Istruzione, dell'Università e della Ricerca (MIUR) within the framework of PRIN2015 project "Clean Heating and Cooling Technologies for an Energy Efficient Smart Grid", Prot. 2015M8S2PA.

Conflicts of Interest: The authors declare no conflict of interest.

Nomenclature

BAT	Best available technology
CNG	Compressed natural gas
DCS	District cooling system
EU	European Union
FCU	Fan coil unit
HP	Heat pump
L–CNG	Liquid to compressed natural gas
LNG	Liquefied natural gas
P	Cooling power profile
PCM	Phase change material
RES	Renewable energy source
SETIS	Strategic Energy Technologies Information System
SFH	Single-family house
TES	Thermal energy storage

References

1. Mazzarella, L. Energy retrofit of historic and existing buildings. The legislative and regulatory point of view. *Energy Build.* **2015**, *95*, 23–31. [CrossRef]
2. Jakubcionis, M.; Carlsson, J. Estimation of European Union residential sector space cooling potential. *Energy Policy* **2017**, *101*, 225–235. [CrossRef]
3. Passerini, F.; Sterling, R.; Keane, M.; Klobut, K.; Costa, A. Energy efficiency facets: Innovative district cooling systems. *Entrepreneurship Sustain. Issues* **2017**, *4*, 310–318. [CrossRef]
4. Frederiksen, S.; Werner, S. *District Heating and Cooling*; Studentlitteratur: Lund, Sweden, 2013; Volume 579.
5. Dominković, D.F.; Rashid, K.B.A.; Romagnoli, A.; Pedersen, A.S.; Leong, K.C.; Krajačić, G.; Duić, N. Potential of district cooling in hot and humid climates. *Appl. Energy* **2017**, *208*, 49–61. [CrossRef]
6. Chiam, Z.; Easwaran, A.; Mouquet, D.; Fazlollahi, S.; Millás, J.V. A hierarchical framework for holistic optimization of the operations of district cooling systems. *Appl. Energy* **2019**, *239*, 23–40. [CrossRef]
7. Vandermeulen, A.; van der Heijde, B.; Helsen, L. Controlling district heating and cooling networks to unlock flexibility: A review. *Energy* **2018**, *151*, 103–115. [CrossRef]
8. Energy Efficiency Directive. Available online: https://ec.europa.eu/energy/en/topics/energy-efficiency/energy-efficiency-directive (accessed on 19 June 2019).
9. Strategic Energy Technologies Information System. Available online: https://setis.ec.europa.eu/ (accessed on 19 June 2019).
10. Connolly, D.; Lund, H.; Mathiesen, B.V.; Werner, S.; Möller, B.; Persson, U.; Nielsen, S. Heat Roadmap Europe: Combining district heating with heat savings to decarbonise the EU energy system. *Energy Policy* **2014**, *65*, 475–489. [CrossRef]
11. David, A.; Mathiesen, B.V.; Averfalk, H.; Werner, S.; Lund, H. Heat Roadmap Europe: Large-scale electric heat pumps in district heating systems. *Energies* **2017**, *10*, 578. [CrossRef]
12. Werner, S. International review of district heating and cooling. *Energy* **2017**, *137*, 617–631. [CrossRef]
13. Gang, W.; Wang, S.; Xiao, F.; Gao, D.C. District cooling systems: Technology integration, system optimization, challenges and opportunities for applications. *Renew. Sustain. Energy Rev.* **2016**, *53*, 253–264. [CrossRef]
14. Buonomano, A.; Calise, F.; Ferruzzi, G.; Vanoli, L. A novel renewable polygeneration system for hospital buildings: Design, simulation and thermo-economic optimization. *Appl. Therm. Eng.* **2014**, *67*, 43–60. [CrossRef]
15. Lee, K.S. A review on concepts, applications, and models of aquifer thermal energy storage systems. *Energies* **2010**, *3*, 1320–1334. [CrossRef]
16. Paksoy, H.O.; Andersson, O.; Abaci, S.; Evliya, H.; Turgut, B. Heating and cooling of a hospital using solar energy coupled with seasonal thermal energy storage in an aquifer. *Renew. Energy* **2000**, *19*, 117–122. [CrossRef]

17. Udomsri, S.; Martin, A.R.; Martin, V. Thermally driven cooling coupled with municipal solid waste-fired power plant: Application of combined heat, cooling and power in tropical urban areas. *Appl. Energy* **2011**, *88*, 1532–1542. [CrossRef]

18. Shu, H.; Duanmu, L.; Zhang, C.; Zhu, Y. Study on the decision-making of district cooling and heating systems by means of value engineering. *Renew. Energy* **2010**, *35*, 1929–1939. [CrossRef]

19. Chow, T.T.; Au, W.H.; Yau, R.; Cheng, V.; Chan, A.; Fong, K.F. Applying district-cooling technology in Hong Kong. *Appl. Energy* **2014**, *79*, 275–289. [CrossRef]

20. Poeuf, P.; Senejean, B.; Ladaurade, C. District cooling system: The most efficient system for urban applications. In Proceedings of the Sustainable Refrigeration and Heat Pump Technology Conference, Stockholm, Sweden, 13–16 June 2010.

21. Zogg, R.; Roth, K.; Brodrick, J. Lake-source district cooling systems. *ASHRAE J.* **2008**, *50*, 55–57.

22. Reverberi, A.; Borghi, A.D.; Dovì, V. Optimal design of cogeneration systems in industrial plants combined with district heating/cooling and underground thermal energy storage. *Energies* **2011**, *4*, 2151–2165. [CrossRef]

23. Kabalina, N.; Costa, M.; Yang, W.; Martin, A. Production of Synthetic Natural Gas from Refuse-Derived Fuel Gasification for Use in a Polygeneration District Heating and Cooling System. *Energies* **2016**, *9*, 1080. [CrossRef]

24. He, T.; Chong, Z.R.; Zheng, J.; Ju, Y.; Linga, P. LNG cold energy utilization: Prospects and challenges. *Energy* **2019**, *170*, 557–568. [CrossRef]

25. He, T.; Karimi, I.A.; Ju, Y. Review on the design and optimization of natural gas liquefaction processes for onshore and offshore applications. *Chem. Eng. Res. Des.* **2018**, *132*, 89–114. [CrossRef]

26. Franco, A.; Casarosa, C. Thermodynamic and heat transfer analysis of LNG energy recovery for power production. *J. Phys. Conf. Ser.* **2014**. [CrossRef]

27. Arteconi, A.; Polonara, F. LNG as vehicle fuel and the problem of supply: The Italian case study. *Energy Policy* **2013**, *62*, 503–512. [CrossRef]

28. Heisch, P. Liquefied to Compressed Natural Gas Opportunities and Strategies. In Proceedings of the GasShow Exhibition and Conference, Warsaw, Poland, 7–8 March 2012.

29. REF-E and SSLNG Watch. *Monitoring of the LNG Final Uses Market in Italy*; Intermediate Report; First half: Vancouver, BC, Canada, 2019.

30. ConferenzaGNL. New LNG Stations: Q8 in Bergamo, Snam-Tamoil Agreement, ENI Commitments. (Nuove Stazioni GNL: Q8 a Bergamo, Accordo Snam-Tamoil, Impegni ENI). Available online: http://www.conferenzagnl.com/nuove-stazioni-gnl-q8-snam-tamoil-eni/ (accessed on 29 July 2019).

31. Franci, T. The prospects for the Italian LNG end-use market in Italy. In Proceedings of the ExpoGNL, Napoli, Italy, 10–11 May 2017.

32. Farzaneh-Gord, M.; Saadat-Targhi, M.; Khadem, J. Selecting optimal volume ratio of reservoir tanks in CNG refueling station with multi-line storage system. *Int. J. Hydrog. Energy* **2016**, *41*, 23109–23119. [CrossRef]

33. Klein, S.; Beckman, A.; Mitchell, W.; Duffie, A. *TRNSYS 17-A TRansient SYstems Simulation Program*; Solar Energy Laboratory, University of Wisconsin: Madison, WI, USA, 2011.

34. Corrado, V.; Ballarini, I.; Corgnati, S.P.; Talà, N. Building Typology Brochure–Italy. In *Fascicolo sulla Tipologia Edilizia Italiana*; Politecnico di Torino, Dipartimento di Energia: Turin, Italy, 2011; pp. 1–117.

35. UNI (Ente Italiano di Unificazione). *Energy Performance of Buildings. Part 1: Evaluation of Energy Need for Space Heating and Cooling*; UNI/TS 11300-1; UNI: Milan, Italy, 2014.

36. UNI (Ente Italiano di Unificazione). *Ergonomics of the Thermal Environment—Analytical Determination and Interpretation of Thermal Comfort Using Calculation of the PMV and PPD Indices and Local Thermal Comfort Criteria. ICS 13.180*; UNI EN ISO 7730:2006; UNI: Milan, Italy, 2006.

37. Pizzetti, C. *Condizionamento Dell'aria e Refrigerazione*; Ambrosiana: Milano, Italy, 2012; Volume 1 e2.

38. American Society of Heating, Refrigerating and Air-Conditioning Engineers (ASHRAE). *ASHRAE Handbook Fundamentals*; 1791 Tullie Circle, N.E.: Atlanta, GA, USA, 2005.

39. UNI (Ente Italiano di Unificazione). *Heating and Cooling of Buildings-Climatic Data-Part 2: Data for Design Load*; UNI/TR 10349-2; UNI: Milan, Italy, 2016.

40. Doninelli, M. I Circuiti e i Terminali Degli Impianti di Climatizzazione (the Circuits and the Terminals of the Air Conditioning Systems). Qauderni Caleffi. Available online: https://www.caleffi.com/sites/default/files/file/quaderno_2_it.pdf (accessed on 29 July 2019).

41. Viessmann. *Air to Water Heat Pump with DC Inverter Technology for Heating and Cooling*; VITOCAL 200-S; Viessmann: Allendorf, Germany, 2010.

42. System with Pre-Insulated Pipes for District Heating & Cooling (DHC). Version 01. May 2014. Available online: http://www.etasuisse.com/assets/Etasuisse%20AG%20-%20Isotechnik%20Katalog%20EN.pdf (accessed on 5 August 2019).

Article

Ground-Source Heat Pumps with Horizontal Heat Exchangers for Space Cooling in the Hot Tropical Climate of Thailand

Arif Widiatmojo [1,*], Sasimook Chokchai [2], Isao Takashima [3], Yohei Uchida [1], Kasumi Yasukawa [1], Srilert Chotpantarat [2,4,5] and Punya Charusiri [2,6]

[1]   Renewable Research Center, National Institute of Advanced Industrial Science and Technology, Japan 2-2-9 Machiikedai, Koriyama-shi, Fukushima-ken 963-0298, Japan; uchida-y@aist.go.jp (Y.U.); kasumi-yasukawa@aist.go.jp (K.Y.)

[2]   Department of Geology, Faculty of Science, Chulalongkorn University 254 Phayathai Rd, Patumwan, Bangkok 10330, Thailand; ps.sasimook@gmail.com (S.C.); csrilert@gmail.com (S.C.); Punya.C@chula.ac.th (P.C.)

[3]   Mining Museum, Akita University, 43 Tegatahebino, Akita-shi, Akita-ken 010-0851, Japan; takashima@gl.itb.ac.id

[4]   Research Program on Controls of Hazardous Contaminants in Raw Water Resources for Water Scarcity Resilience, Center of Excellence on Hazardous Substance Management (HSM), Chulalongkorn University, Bangkok 10330, Thailand

[5]   Research Unit Control of Emerging Micropollutants in Environment, Chulalongkorn University, Bangkok 10330, Thailand

[6]   Department of Mineral Resources (DMR) King Rama VI Road, Ratchatewi 10400, Bangkok

*   Correspondence: arif.widiatmojo@aist.go.jp; Tel.: +81-29-861-0529

Received: 25 February 2019; Accepted: 29 March 2019; Published: 2 April 2019

Abstract: The cooling of spaces in tropical regions, such as Southeast Asia, consumes a lot of energy. Additionally, rapid population and economic growth are resulting in an increasing demand for space cooling. The ground-source heat pump has been proven a reliable, cost-effective, safe, and environmentally-friendly alternative for cooling and heating spaces in various countries. In tropical countries, the presumption that the ground-source heat pump may not provide better thermal performance than the normal air-source heat pump arises because the difference between ground and atmospheric temperatures is essentially low. This paper reports the potential use of a ground-source heat pump with horizontal heat exchangers in a tropical country—Thailand. Daily operational data of two ground-source heat pumps and an air-source heat pump during a two-month operation are analyzed and compared. Life cycle cost analysis and CO_2 emission estimation are adopted to evaluate the economic value of ground-source heat pump investment and potential CO_2 reduction through the use of ground-source heat pumps, in comparison with the case for air-source heat pumps. It was found that the ground-source heat pumps consume 17.1% and 18.4% less electricity than the air-source heat pump during this period. Local production of heat pumps and heat exchangers, as well as rapid regional economic growth, can be positive factors for future ground-source heat pump application, not only in Thailand but also southeast Asian countries.

Keywords: ground source heat pump; tropical climate; horizontal heat exchanger

1. Introduction

Southeast Asian countries have experienced rapid economic growth, at an average rate of 5.2% per year since 2000. The rapid growth has been followed by an increase in the energy demand. In 2016, the total primary energy consumption in the region reached 643 million tons of oil equivalent. While

Southeast Asian countries may not be considered a major global CO_2 contributor, the CO_2 emissions of those countries rose from 711 MT in 2000 to 1288 MT in 2015 [1,2]. The total generation of electricity in the region increased from 370 TWh in 2000 to 868 TWh in 2015. By 2015, 83.4% of electricity was generated by burning fossil fuels (i.e., coal, natural gas, and oil). Serious action must therefore be taken immediately, in order to reduce the fossil fuel dependency [3].

Thailand accounts for 21.7% of the primary energy demand in Southeast Asia [1]. By 2015, the national primary energy consumption and total electricity generation were respectively 135 million tons of oil equivalent and 178 TWh, with fossil fuel accounting for 80.7% and 91.6%, respectively. In 2017, the generation of electricity emitted 96.035 MT of CO_2 [2]. Air conditioners consume much of a household's electricity demand. According to a report published by The Japan Refrigeration and Air Conditioning Industry Association, Thailand's domestic total air conditioner demand in 2016 was 1.56 million units, the third largest demand in southeast Asia after Indonesia and Vietnam [4]. Data published by the Ministry of Energy of Thailand suggest that the residential sector consumed 20.4% of national electricity, 46% and 17% of which were used for air conditioning and refrigeration, respectively [2]. These sectors have high potential energy savings [2,5,6]. Governments of Southeast Asian countries are aware of this problem, and are thus considering several actions that promote the higher efficiency of air conditioners in this region [7,8]. The residential energy growth in Thailand is greatly determined by the increasing number of households, as well as an increasing income per capita. The use of energy-efficient products is an important way of restraining the household energy demand. However, the market prices of energy-efficient products, such as five-star-rated energy-saving air conditioners, tend to be higher than those of regular products [9,10].

Meanwhile, it has been shown that the increase in the energy demand in Thailand accelerate climate change and the urban heat island (UHI) phenomena [11–13]. Arifwidodo and Chandrasiri have estimated that the annual increase of average temperature in an urban area (Bangkok) and a sub-urban area (Pathumthani) are increasing as a result of UHI. The UHI severity index has been found to be higher than those of other major cities in the world, such as Shanghai, San Diego, and San Francisco, and within a similar range as Tokyo [13]. On the other hand, other studies have found possible countermeasures of UHI in the Tokyo area by utilizing a district heating-cooling system and ground-source heat pumps (GSHPs) [14,15].

The government of Thailand begun to improve the energy efficiency of air conditioners in the 1990s, and started labeling the ratings of air conditioners by the mid-1990s. End consumers are thus supposed to be well aware of the labeling system. The efficiency of air conditioners was improved by the introduction of an inverter, although its market penetration rate and market share remain low. Meanwhile, the high demand for air conditioning units has offset efficiency improvements [6].

Among various alternative energy sources, the ground-source heat pump (GSHP), which utilizes a relatively constant ground temperature, is widely applied for the cooling and heating of spaces. Instead of exchanging heat with the outdoor environment, as in the case of a normal air-source heat pump (ASHP), the GSHP uses the ground as a heat sink (for the cooling of spaces) and a heat source (for the heating of spaces).

GSHP systems are generally classified as open-loop and closed-loop systems. Closed-loop systems can be further classified into systems having vertical and horizontal arrangements of the ground heat exchanger (GHE). The vertical closed-loop system has higher thermal efficiency, and the heat transfer rate can be further improved through the convective heat transfer of groundwater flow. Although the required site area for this system is small, the system has a high initial cost for drilling and installation of the GHE. Meanwhile, the horizontal (shallow) closed-loop system is relatively inexpensive, as it requires no vertical borehole and just shallow trenches that can be dug through manual (human) labor or mechanical means, such as the use of a mini-excavator [16–18]. However, this system requires a larger footprint for installation. The drilling cost of a vertical GHE in Japan is USD 6700 for a 50 m borehole, while the same borehole costs USD 3000 in Thailand. In Thailand, meanwhile, the average wage for manual (human) labor is around USD 13 per day. The horizontal system thus has a great advantage in terms of the initial cost.

In most cases, however, the thermal performance of the horizontal closed system is lower than that of the vertical closed system, as the soil temperature at a shallow depth fluctuates and is strongly affected by the ambient temperature and near-surface heat flux [18]. Thus, for a high cooling load demand, such as in the case of the central cooling of an office or public building with a GSHP, a horizontal heat exchanger may not be adequate.

The horizontal closed system is increasingly being studied. Several studies have compared shallow linear, helical, and slinky GHEs in numerical simulation, and have found that the helical configuration has the best performance [16,17,19]. They have also found that the thermal conductivity surrounding a GHE and the flowrate of the heat transfer fluids are the most important parameters. Fujii et al. performed numerical simulations of slinky GHEs installed at different depths, simplifying the GHEs as thin flat plates [20]. Their simulation results agreed well with experimental data. In subsequent work, they presented the results of a slinky GHE field test during heating (winter) and numerical simulations of double- and single-layer arrangements, and found that the double layer has a lower energy cost per unit of site area, owing to its better performance [21]. Recent studies have also remarked the importance of soil properties, moisture content, environmental parameters, and installation design to heat-pump performance [22–26].

Recent research on GSHPs also has focused on the hybrid system. The GSHP hybrid system incorporates another thermal system, i.e., a desiccant or solar thermal energy. The hybrid GSHP–desiccant system allows better and more effective means of controlling space humidity and air temperature [27–29]. By taking direct solar heat energy, the efficiency of a GSHP during heating can be significantly improved. The hybrid GSHP–solar system offers higher thermal performance for applications, such as water heating, heat storage, or drying [30–33].

Unlike the case of most GSHP applications in a four-season climate, the cooling load is predominant in a tropical climate. The application of the GSHP in a tropical climate thus uses the ground mainly as a heat sink to remove heat from a building. Furthermore, the difference between ground and air temperatures is negligible.

To the extent of our knowledge, only few studies have focused on the application of the GSHP in tropical climates. One study showed that GSHP are expected to replace underperforming air-cooled condensers in Singapore [34]. Permchart and Tanatvanit used the ground as a heat sink for direct-expansion GSHP, by directly burying the refrigerant piping exiting the compressor of the heat pump [35]. Yasukawa et al. identified regional variation in the subsurface temperature by investigating the vertical temperature variation in several observation wells around the Chao Phraya Plain in Thailand and the Red River Plain in Vietnam [36]. In several areas, they observed subsurface temperatures lower than the monthly mean air temperature, while in other areas, subsurface temperatures were higher, but still lower than the monthly maximum air temperature. Furthermore, the authors emphasized that the application of GSHPs in these areas could take advantage of advective heat transfer due to groundwater flow. Following their study, a GSHP system was installed in Kamphaengphet, Thailand [37]. The system uses a single 56 m borehole with a double U-tube GHE. Long-term performance results show that an average coefficient of performance (CoP) of 3 can be achieved. Additionally, during successive operation, the borehole temperature increased, but recovered to its initial temperature after a week, and there was no long-term increase in the subsurface temperature after more than a year of operation. Uchida et al. conducted subsurface groundwater surveys and a stable isotope evaluation on the Chao Phraya Plain [38]. Their results show differences in the subsurface thermal gradient between lower and upper plains, due to thermal conduction by regional groundwater flow. Furthermore, they remarked that GSHPs installed in areas with different groundwater thermal conduction characteristics may have different performance efficiencies. The most recently published research on the application of GSHPs in Bangkok with vertical boreholes and a single U-tube configuration showed advantages in terms of energy savings compared with a normal ASHP [39].

Various economic analyses can be applied to assess the economic value of GSHPs—e.g., present value (PV) analysis, internal rate-of-return analysis, net-benefit analysis, payback analysis, and benefit-to-cost ratio analysis [40]. Noorollahi et al. performed a numerical simulation and economic evaluation using the PV for the application of GSHPs to the energy supply of greenhouses in Iran [41]. The annual cost has been used to evaluate the feasibility of GSHP application in Turkey [42]; it was concluded that the GSHP system is economically preferable to the ASHP system for cooling purposes. Esen et al. [43] conducted a techno-economic assessment of GSHPs for heating in Turkey. Go et al. [44] evaluated the economic feasibility of various configurations of the shallow spiral coil loop heat exchanger, using the PV, internal rate of return, and savings-to-investment ratio. Zu et al. analyzed the economic application of GSHPs in hot and humid climates using the PV [45].

Owing to aforementioned concerns, the objectives of present work are (i) to demonstrate the applicability of GSHPs, using shallow heat exchangers compared to the ASHP, through the experimental results in the hot tropical climate of Thailand; and (ii) to highlight important financial considerations by analyzing the prevailing factors that must be considered in order to make GSHP application in Thailand and other Southeast Asian countries economically attractive.

2. Climate of Thailand

Located in a tropical area, Thailand has a topography that can be divided into five areas and two regions. The upper (higher latitude) region is comprised of northern, northeastern, central, and eastern areas, while the lower region is comprised of the southern area. According to the Koppen climate classification, the upper region has a tropical monsoon climate, while the lower region has a tropical savanna/wet climate, with an equatorial climate in a small part of southern Thailand. The year is divided into three seasons. The rainy season of the southwest monsoon is from mid-May to mid-October, with the rainfall being highest from August to September. Winter of the northeast monsoon is from mid-October to mid-February. The summer or pre-monsoon season is from mid-February to mid-May.

In contrast with the upper region, which has high temperatures and a long warm period, the lower region has milder temperatures and higher humidity, as well as less diurnal and seasonal temperature variation [46]. Figure 1 shows the daily average, minimum and maximum temperatures, and average relative humidity recorded in 2017 at two measuring stations located in Bangkok (13°45′09″ N) and Lopburi (14°48′00″ N), ±150 km northeast of Bangkok. The recordings indicate that there are only small seasonal temperature variations, with the temperatures being highest during April–May. Note that Bangkok has slightly lower annual temperatures but higher humidity than Lopburi, as Bangkok is located at a lower latitude and close to the Gulf of Thailand. The green shading in the figure indicates average underground temperatures recorded in observation wells [36]. It is also important to note that there is almost no difference between the annual average temperature and ground temperature. However, Bangkok has a slightly higher underground temperatures with a wider range compared with Lopburi.

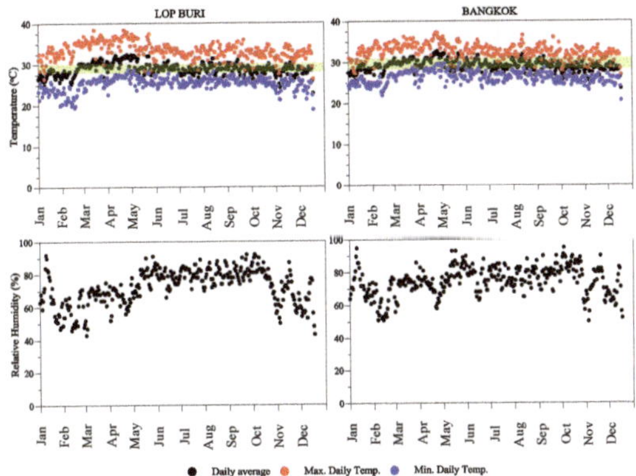

Figure 1. Annual temperature (**upper**) and relative humidity (**bottom**) variations measured in Bangkok and Lopburi Province (2017). Green shading shows the average shallow ground temperature at depths of 0–100 m [36].

3. Configuration of the Heat Pump System

3.1. Shallow Ground Heat Exchanger

Shallow horizontal heat exchangers were installed at the Saraburi Campus of Chulalongkorn University. The campus (14°31′17.4″N 101°1′07.1″E) is located in Saraburi province, northeast of Bangkok, as shown in Figure 2. The upper soil is comprised of mainly volcanic-derived, clayey-sandy soil.

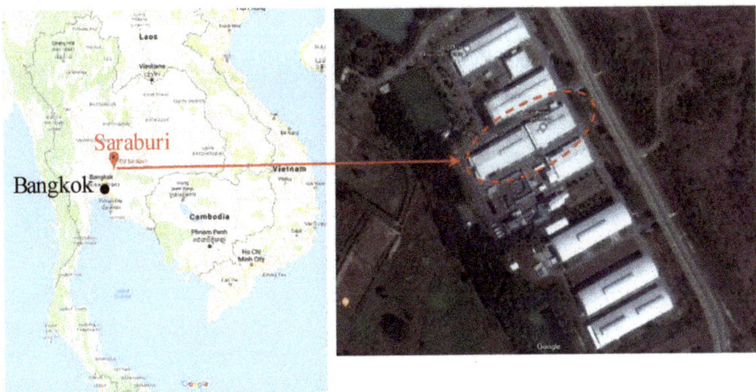

Figure 2. Location showing the installation site of ground-source heat pump (GSHP) systems at the Saraburi campus of Chulalongkorn University, Saraburi Province.

There were four groups of heat exchangers, as shown in Figure 3. Two types of heat exchanger were used, namely sheet-type (carpet-type) and high-density polyethylene (HDPE) pipe GHEs. The carpet-type heat exchanger had overall dimensions of 5.6 m × 0.9 m and is comprised of 117 small HDPE tubes (with an outer diameter of 6 mm), as shown in Figure 4. HDPE pipes with a diameter of 3.2 cm and thickness of 2.4 mm were used in both slinky and helical arrangements, as shown in Figures 5 and 6. In total, 500 m of HDPE pipe and two carpet-type heat exchangers were installed over a footprint of 73.8 m². The ground heat exchangers were not influenced by the groundwater. The detailed setups of GHEs were as follows.

In Group 1, two layers of 100 m of HDPE slinky pipes were installed in the bottom of a trench having dimensions of 14 m × 2 m × 1.5 m (depth), with a vertical separation of 50 cm. Figure 5 shows the preparation and installation of the Group 1 heat exchanger.

In Group 2, two layers of 50 m of HDPE slinky pipes were installed in the bottom of a trench having dimensions of 4 m × 2 m × 1.5 m (depth), with a separation of 50 cm.

In Group 3, two layers of GHEs were installed in the bottom of a trench having dimensions of 20 m × 2 m × 0.8 m (depth). The bottom layer was comprised of two carpet-type heat exchangers connected in a series, while a slinky-pipe heat exchanger was installed 0.3 m above the bottom layer. Figure 4 shows the installation of the carpet-type heat exchanger.

Lastly, in Group 4, two 50-m HDPE pipes in a helical configuration were installed in the bottom of two trenches having dimensions of 3 m × 1.3 m × 2 m (depth). The two pipes were connected in a series. Figure 6 shows the installation of this heat exchanger.

3.2. Room Cooling Experiment

An experimental room (having dimensions of 3 m × 8 m × 2 m) in Building 1 of the Center of Fuels and Energy was used for an operational cooling test. The material of the building's wall was fiber cement board having thermal conductivity of $\lambda = 0.14$ Wm^{-1}K^{-1} and heat capacity $C = 600$ Jkg^{-1}K^{-1} [47]. Three fan coil units (FCUs)—one ASHP and two GSHP (GSHP 1 and GSHP 2) systems—were installed. Both GSHPs use antifreeze solution (40% propylene glycol) as ground heat transfer fluid. The antifreeze solution was used in order to reduce the risk of corrosion. GSHP 1 was a Japanese GSHP with reversible functions for cooling and heating. GSHP 2 was an ASHP, modified by changing the original heat exchanger with a plate heat exchanger (Kaori-K050X22, 7.03 kW) to allow heat exchange between the R410A refrigerant fluid and ground-loop circulation fluid. Figure 7 shows the heat exchanger replacement work of the modified GSHP. Specifications for each system are presented in Table 1. Both GSHPs were connected in a series to GHEs, as shown in Figure 3. A Graphtec GL240 logger recorded data from the power meter and thermocouples that measured the outdoor, indoor, and circulation-fluid inlet and outlet temperatures at 10-min intervals. For the ASHP, only the power consumption and outdoor and indoor temperatures were recorded. The thermocouples used for measurements are described in Table 2.

Table 1. Technical specifications of the heat pumps used in the present study.

Heat Pump	Cooling Capacity (kW)	Heating Capacity (kW)	Inverter	Refrigerant	Energy Rating[1]	Remarks
GSHP 1	4	5	Yes	R410a	-	Imported from Japan
GSHP 2	N/A[2]	-	No	R410a	N/A[3]	Replaced with 7.03 kW Water-R410A plate heat exchanger
ASHP	3.5	-	Yes	R410a	5 stars	-

[1] Certification standard issued by Electricity Generating Authority of Thailand (EGAT); [2] the original Cooling capacity was 3.6kW; [3] the original energy rating was five stars (2011 standard certification).

Table 2. Temperature sensor specifications.

Sensor Location	Sensor Type	Accuracy
Indoor air temp.	T type thermocouple	+/−0.5 °C
Atmospheric air temp.	T type thermocouple	+/−0.5 °C
Heat transfer fluid inlet	Pt100 platinum resistance	+/−0.2 °C
Heat transfer fluid outlet	Pt-100 platinum resistance	+/−0.2 °C
Ground Temperatures	NTC thermistor	+/−0.2 °C

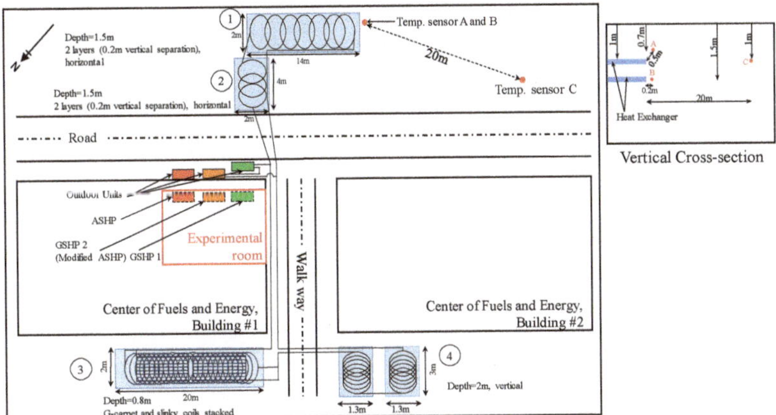

Figure 3. Schematic diagram of the connection of heat pumps and GHEs; the number in the circle represents the grouping of the ground heat exchangers.

4. Data Analysis

The CoP is the ratio of heat removed from the room (see Figure 8) to the work required:

$$CoP = \frac{Q_C}{W_T} \tag{1}$$

Here, Q_c (in W) is the rate of heat removed from the building and W_T (in W) is the total electrical power consumption, calculated as

$$W_T = W_C + W_F + W_P \tag{2}$$

where W_C, W_F, and W_P are the electrical power for the compressor, fan, and circulation pump, respectively.

Figure 4. Carpet-type heat exchanger installation.

Figure 5. Installation of high-density polyethylene (HDPE) slinky pipes (Group 1).

Figure 6. Installation of a helically configured heat exchanger (Group 4).

Figure 7. Modification of the air-source heat pump (ASHP) and installation of the plate heat exchanger.

As the heat pump performance data record the temperature and flowrate of GHE fluid inlet and outlet from the GSHP, Equation (1) can be rewritten as

$$\text{CoP} = \frac{Q_C}{W_T} = \frac{Q_H - W_C}{W_T} \tag{3}$$

Here, Q_H (in W) is the rate of heat rejection into the ground, expressed as

$$Q_H = (T_{out} - T_{in})\rho c V_m \tag{4}$$

where T_{out} and T_{in} (in C) are, respectively, the heat exchange fluid temperatures at the GSHP outlet and inlet, while ρ (kg/m^3), c (J/(kgC), and V_m (m^3/s) are, respectively, the density, specific heat capacity, and flowrate of the heat exchange fluid.

The variation of data within the operational period of each heat pump is quantified in simple standard deviation analysis as

$$\sigma = \sqrt{\frac{\sum_{i=1}^{N}(x_i - \overline{x})^2}{N-1}}, \tag{5}$$

where x_i is the observed value, $\overline{x}$ is the average value, and N is the number of data points.

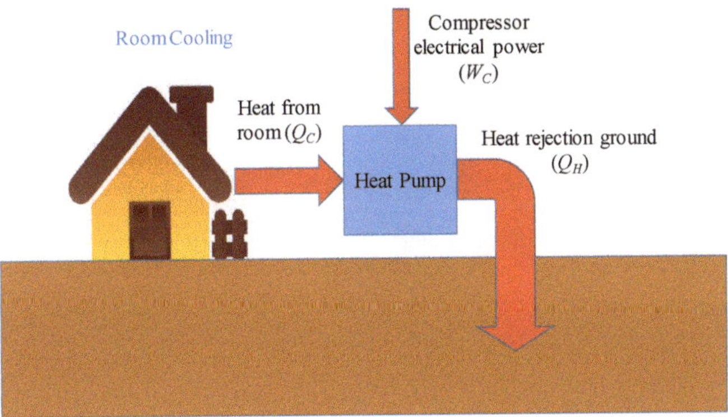

Figure 8. Illustration of energy flow during cooling.

5. Results and Discussion

5.1. Experiment and Analysis

The heat pump performance data presented in this paper were recorded during May–June 2018. The heat pumps were used for intermittent cooling during the operational period.

Figure 9 shows that the temperature was highest in April, while May and June are among the hottest months of the year. Meanwhile, the rainfall and humidity were highest in September. This suggests that the thermal loads of heat pumps vary through the year. There was a high sensible cooling load around April, and a high latent cooling load around September.

From May to June 2018, GSHP 1, GSHP 2, and the ASHP respectively, operated for 19, 18, and 17 days, with the daily average operational period being 9.6 hours; it should be noted that the start–stop schedule was not fixed at an exact time. The experiment was conducted on a daily basis during weekdays and sometimes during the weekend, in case the experimental room was being used. Apart from the automated logging system, personnel who turned the heat pumps on and off wrote down the start/stop time, as well as the analog energy meter readings in a logbook. The data gathered from the data logger were then cross-validated against the logbook data. The room temperature was set at a constant 25 °C for all heat pumps. The average heat exchange circulation flowrate of GSHP

2 was 11.56 L/min (= 1.92×10^{-4} m^3/s), owing to the low-capacity circulation pump, in contrast to 22.27 L/min (= 3.71×10^{-4} m^3/s) for GSHP 1. Figure 10 shows the variation in the outdoor temperature during the operational time. Average operational temperatures were 32.4 and 31.9 °C for May and June, respectively.

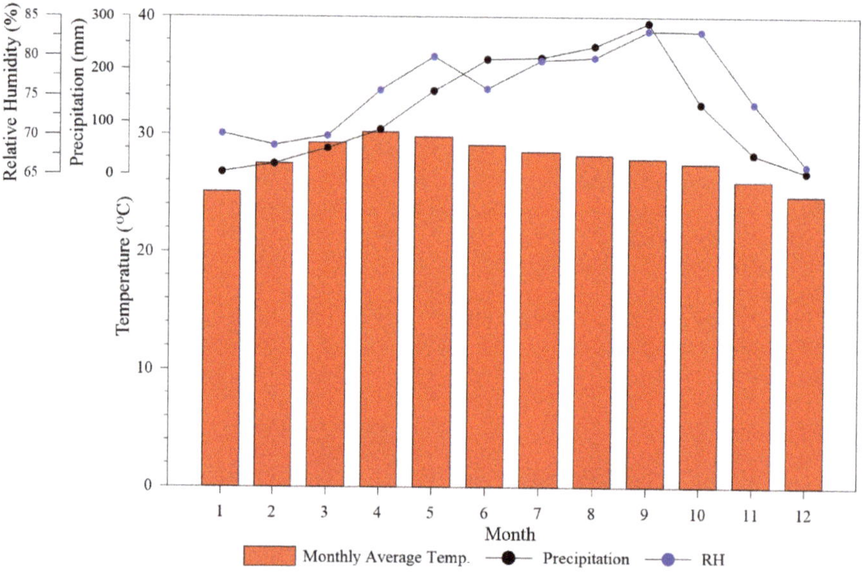

Figure 9. Annual variations of temperature, precipitation, and relative humidity at Saraburi.

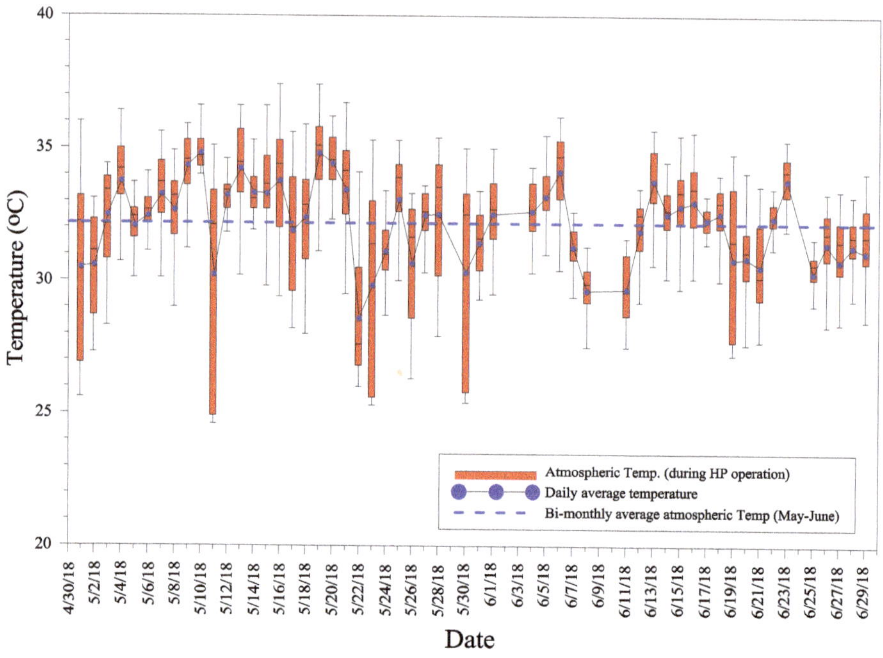

Figure 10. Atmospheric temperature variation during the heat pump operation.

The logged temperatures and power consumption during the operation of GSHP 1, GSHP 2, and the ASHP are presented in Figures 11–13. In addition to ASHP data, calculations of the CoP and heat rejection rate are shown. A comparison of the performances of GSHP 1 and GSHP 2 reveals that even though the circulation flow rate of GSHP 1 was half that of GSHP 2, there was no appreciable difference in the heat rejection rate—i.e., 4.26 kW for GSHP 1 versus 4.29 kW for GSHP 2. This is attributed to differences in the inlet and outlet temperatures of the GSHPs.

Further analysis indicates that for two months of operation, during which GSHPs were used for 37 days, there was no increase in the daily final inlet temperature. This suggests that there was no long-term rise in the background temperature, at least during the experimental period. Different inlet and outlet temperatures on GSHP operational days were simply due to different cooling loads resulting from variations in the outdoor temperature.

GSHP 1 and the ASHP showed scattered (wide range) values of power consumption, compared with steady values for GSHP 2. This was due to the ability of the inverter to regulate the compressor speed and precisely control the evaporator outlet temperature [48]. The variations can be explained by evaluating the standard deviations of the room temperature, power consumption, and heat rejection rate.

Figure 14 compares the performances of all heat pumps. Each error bar represents the standard deviation of data. The standard deviations of the power consumptions of GSHP 1 and the ASHP were larger, indicating larger variations, while the temperature data of GSHP 1 and the ASHP had smaller standard deviations in comparison with the standard deviations for GSHP 2. A comparison of the average electrical consumption shows that the GSHPs required less electrical input than the ASHP. The average electrical consumptions of GSHP 1 and GSHP 2 were 658.7 and 648.6 W, respectively, being 17.1% and 18.4% less, respectively, than the electrical consumption of 795.5 W for the ASHP.

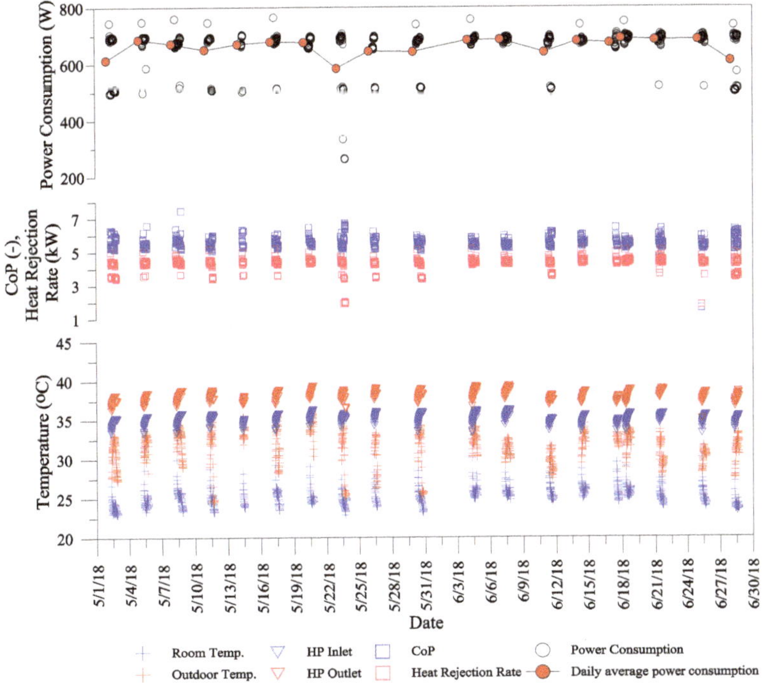

Figure 11. Performance of GSHP 1 (10-min interval data during the operational period).

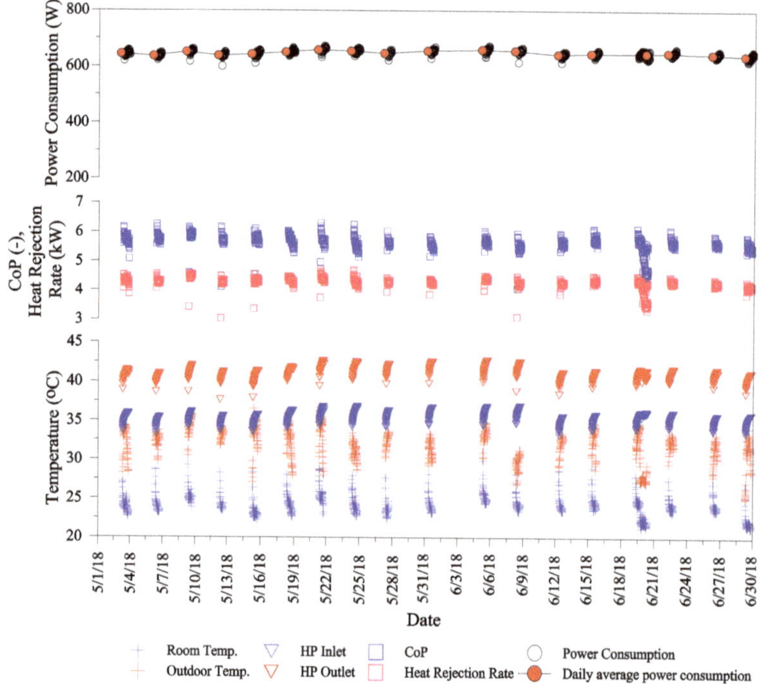

Figure 12. Performance of GSHP 2 (10-min interval data during the operational period).

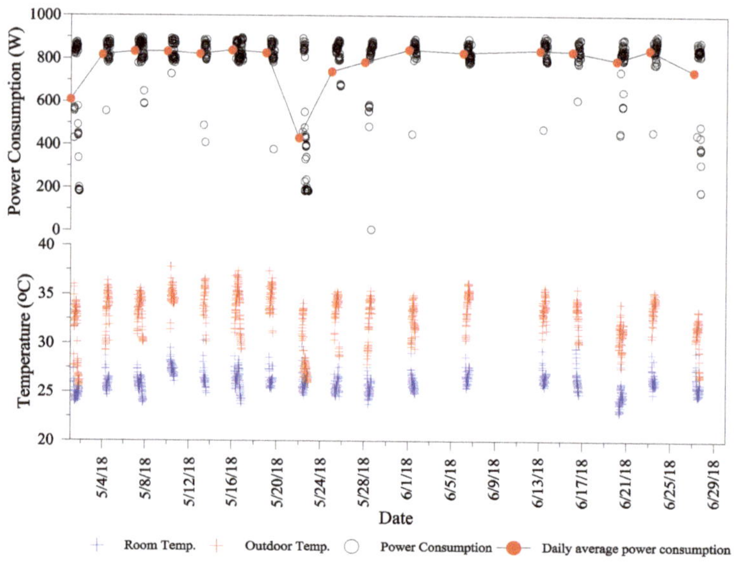

Figure 13. Performance of the ASHP (10-min interval data during the operational period).

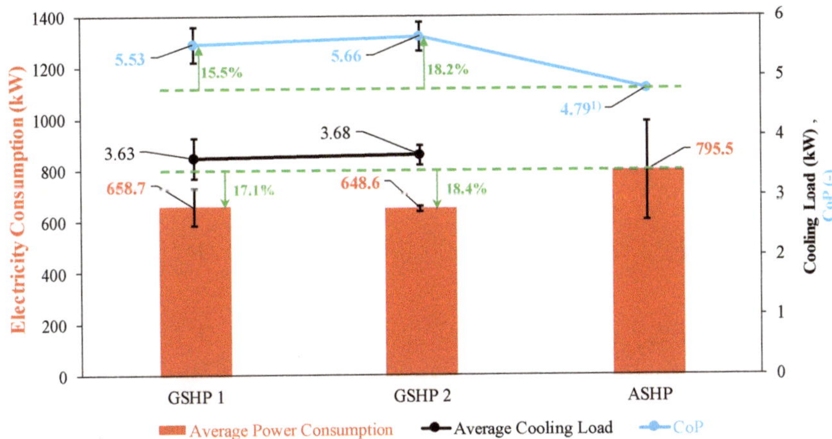

Figure 14. Performance comparison of the GSHPs and ASHP; the coefficient of performance (CoP) of the ASHP is calculated assuming an average cooling load similar to that of GSHP 1 and GSHP 2.

The ASHP consumed more electricity than the GSHPs, with most of the data extending within 800 W or more. However, when the average outdoor temperature was low (e.g., see May 1, May 22, and June 27 in Figures 10 and 13), the ASHP operated with low electric power, owing to the low cooling load and inverter control. The calculated cooling loads of the two GSHPs had similar values despite having different standard deviations. Furthermore, supposing that the average cooling load during the two-month period did not change greatly, the cooling load calculated for a GSHP (3.65 W) was used to estimate the CoP of the ASHP, as shown in Figure 14.

Figure 15 compares temperature data of GSHP 1 and GSHP 2 for similar operational outdoor temperatures (see also Figure 10) on May 20 and May 9, respectively. GSHP 1 operated for 8.33 hours, while GSHP 2 operated for 9 hours. GSHP 2 began operation earlier in the morning, while GSHP 1 stopped operation later in the evening. This can be seen from the low outdoor temperature of GSHP 2 at the beginning of operation, in contrast to the low outdoor temperature of GSHP 1 at the end of operation. The results reveal that the outlet temperature of GSHP 2 was higher than that of GSHP 1, although the operating conditions were nearly the same. The inlet and outlet temperatures of the two systems rose over time, following a similar linear trend. It is interesting that even though the outlet temperature of GSHP 2 was roughly 5 °C higher than that of GSHP 1, the difference in average power consumption for the two heat pumps was not great. From the current standpoint, bearing in mind that the current GHE circulation flow rate was low, a higher flow rate may be considered to increase the performance of GSHP 2.

Figure 16 shows the performance of the ASHP. During operation at a high outdoor temperature, the ASHP required a power input higher than the average power consumption of GSHP 1 and GSHP 2. However, as soon as the outdoor temperature dropped below 27.5 °C, after about 7 hours of operation, the power consumption fell below the average value for GSHP 1 and GSHP 2. This temperature range is a turning point, where one may consider that the application of a GSHP has no advantage over the application of an ASHP. There is a close relationship between thermal comfort and occupants' psychological and physical considerations [49–51]. The personal comfort zone may thus be different from the standard definition of thermal comfort [5,8,51,52]. Studies focusing on indoor thermal comfort found that people living in the tropics tend to have an acceptable temperature comfort zone higher than that of ASHRAE standard 55 [5,50,52–56]. Thus, at this temperature, people may require only the circulation of air through the use of a fan to obtain an acceptable comfort level.

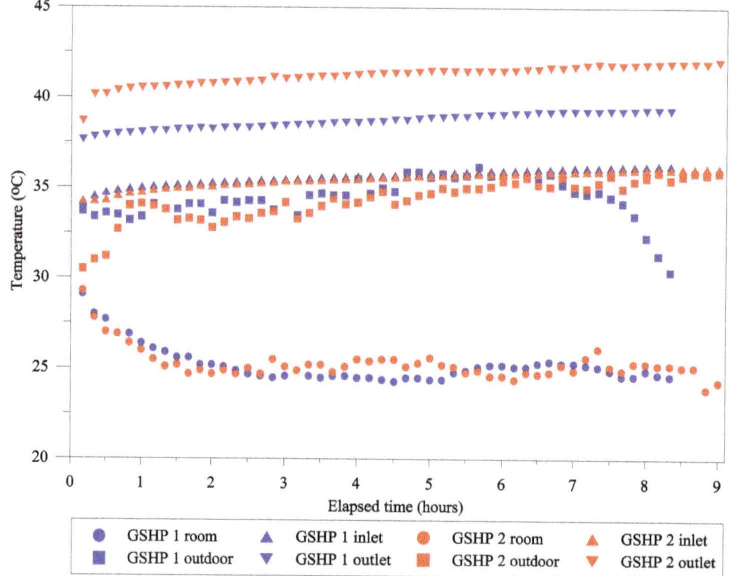

Figure 15. Temperature data of GSHP 1 (May 20) and GSHP 2 (May 9).

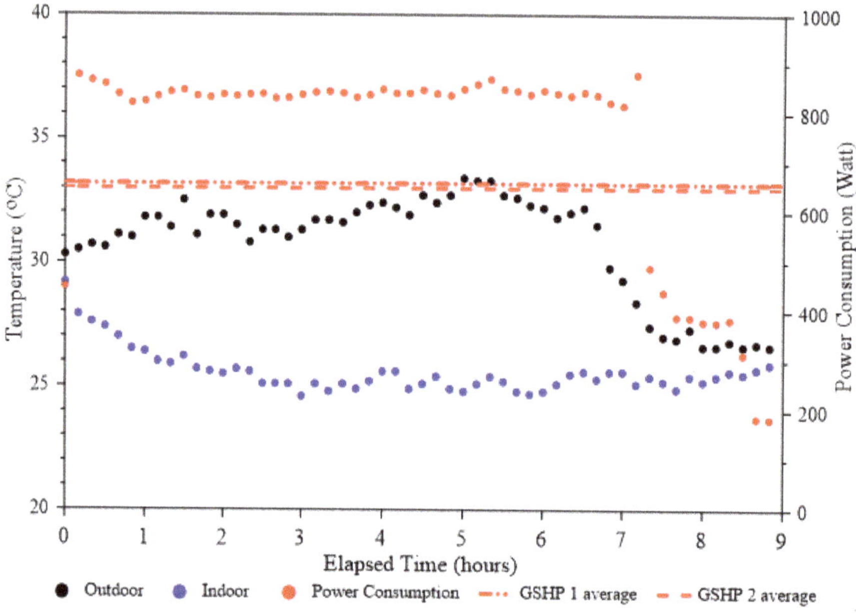

Figure 16. One-day operational data of the ASHP (June 27).

5.2. Ground Temperature Data

One of the most important parameters that affects the performance of GSHP is the soil temperature. It fluctuates due to both the heat transfer from/to the heat exchangers and the natural heat transfer from the surrounding environment [57–62].

Three ground temperature sensors were installed in the experimental site. Sensor A and B were installed at depths of 0.7 m and 1.5 m, respectively, in a single vertical hole, close to the end of the Group 1 heat exchanger (see Figure 3), with a 60-min sampling interval. Sensor C was installed 20 meters away, to measure the background temperature. However, from middle of March 2018 onward, no data were recorded, due to broken underground wiring. Figure 17 shows the recorded temperatures between 1 December 2017 and 13 March 2018. During this period, a similar intermittent experimental pattern was carried out. Broken lines show the daily maximum and minimum temperatures during the same period. The general trend of the ground temperature could be divided into two periods, i.e., the declining ground temperature before January, followed by the increasing temperature after January. The declining temperature can be observed during December as the average air temperature decreases. During this month, the ground temperature fluctuations, as can be observed from sensors A and B, were lower, due to the fact that the thermal load of GSHP was low, resulting in the low heat rejection rate. It is also clear that the temperatures from all three sensors show that the ground temperatures are close to those of the maximum daily temperature. Temperature fluctuation could be observed from all sensors. Fluctuations shown by sensor C indicated that the ground temperature fluctuations at 1 m of depth, due to the daily variation of air temperatures. A sharp fluctuation pattern can be observed from sensor B, due to the increasing ground temperature followed by the declining temperature during the GSHP's operation cycle. However, positioned further away from heat exchanger, less fluctuation is observed at position A.

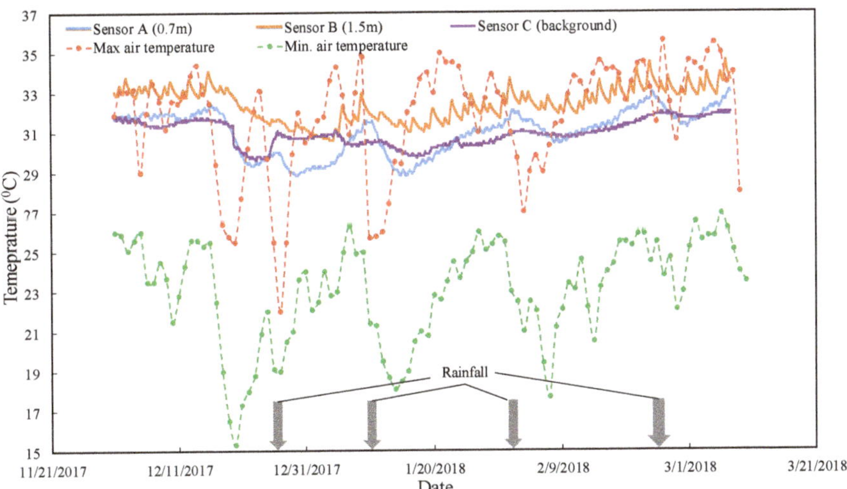

Figure 17. Ground temperatures and maximum–minimum daily air temperatures.

Looking closely at Figures 17 and 18, the temperature pattern of sensor B indicated that the intermittent use of GSHP and ASHP could provide enough of a period for stabilizing ground temperature and preventing excessive ground temperature rise, which could eventually lower the GSHP performance. It is likely that the use of GSHP alone will not provide better thermal performance compared to its present intermittent use with ASHP.

It is interesting to note that the ground temperature at position A was strongly affected by the rainfall infiltration. When rain occurred, the temperature at this position (0.7 m depth) started to decline by about 2–3 °C. The declining temperature at this depth was prolonged for 2–4 days after the rainfall. Figure 18 shows the detail of ground temperatures from February 20–March 3. After the rain occurred in February 23, the temperature at sensor position A started to decline. However, the temperature at the background (position C, 1 m depth) seemed to be unaffected by the rain, except more rain on February 27, in the morning, when a slight temperature decline was observed. In relatively dry conditions, the ground temperature at position B (1.5 m depth) increased during the heat rejection period. In contrast, the soil at position A increased after the heat rejection was finished. A similar pattern also can be observed during the soil desaturation period after the rainfall (February 28). During the soil saturation process (February 23–27) following the rainfall, such an occurrence was not observed.

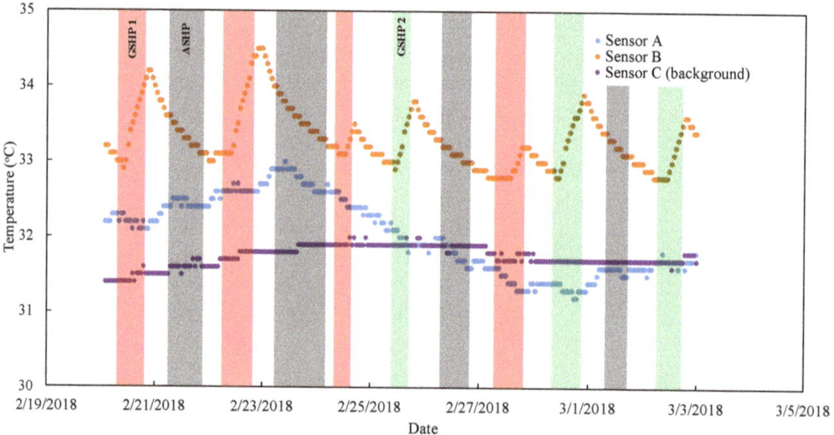

Figure 18. Ground temperatures during heat pump operation (February 20–March, 3).

Rainwater is in thermal equilibrium with the atmospheric temperature when it reaches the ground's surface. It infiltrates the ground surface at a lower temperature, in contrast to the upper ground temperature. The heat transfer at shallow depths involves a complex mechanism, including conduction by soil matrix, convection by liquid water, and sensible and latent heat transfer by water vapor movement [60,63–65]. However, further analysis on heat and moisture transfers at shallow depths and their effect on the performance of GSHPs are not discussed in this paper, due to the absence of the performance data of heat pumps and soil moisture, i.e., soil hydraulic parameters, volumetric water content, degree of saturation, and pore–water pressure. It has been remarked in previous studies that soil thermal conductivity is strongly related to soil water saturation [62,66–69]. Go et al. evaluated the effects of rainfall infiltration on the performance of GSHP, with shallow horizontal heat exchangers via numerical analysis [69]. Their study remarked on the free convective heat transfer at the upper region of a shallow heat exchanger during heat rejection, due to both air density and temperature gradients. They pointed out the relationship between increasing thermal conductivity due to rain infiltration and GSHP thermal performance. Ground heat exchanger performance prediction accuracy is significantly affected by soil moisture transfer and properties [62].

5.3. Life-Cycle Cost Analyses of the Potential Annual CO_2 Reduction

In order to make GSHP application gain attention among people in southeast Asian countries, it is important to understand the economic benefits of its application.

The present study conducts a life-cycle cost analysis using the PV. The PV is defined as the current value of future cash flows:

$$PV_t = \frac{C}{(1+i+j)^t} \quad t = 1,2 \cdots n_T \tag{6}$$

The lifetimes of the heat pumps are assumed to be $n_T = 15$ years, and both the interest rate i and inflation rate j are set at 1.5%. The net present value (NPV) is the total PV for the given period of investment:

$$\text{NPV} = \sum_{t=1}^{n_T} \text{PV}_t \tag{7}$$

Heat pumps are assumed to operate on average for 9 h per day and 320 days per year. Further assumptions are that the land cost can be ignored and that no taxation is imposed. The annual CO_2 emission generated in running the heat pumps is calculated by considering an annual emission intensity (2017) of 0.471 kg/kWh [2]. Under the aforementioned operating period and experimental results, the annual CO_2 emission is 1078.4 kg CO_2/year for the ASHP, and 894.16 and 880 kg CO_2/year for GSHP 1 and GSHP 2, respectively.

The costs used in the life-cycle cost analyses are presented in Table 3. The table shows that the cost of GSHP 1 is 4.3 times that of the ASHP. This is because the Japanese domestic price is used for GSHP 1, as it is sold solely in Japan and was imported to Thailand for the purpose of research.

Table 3. Initial and installation costs of each heat pump and heat exchanger.

Heat Pump	Life-Cycle Cost	Cost[1]
GSHP 1	Total cost	5700
GSHP 2	Total cost	900
Ground Heat Exchanger	Material cost	225
	Installation cost	360
ASHP	Unit cost	630
	Installation cost	90
Others	Electricity cost	0.135[2]

[1] All costs are in USD; [2] Electricity cost is in USD/kWh.

Figure 19 presents the NPV of each heat pump when considering power reductions of 17.1% and 18.4% for GSHP 1 and GSHP 2, respectively, as a base case. The figure reveals that the NPV period for GSHP 2 is equal to that of the ASHP after 15 years, when all systems reach their assumed maximum lifespans. From an economic perspective, GSHP 2 has no advantages over the ASHP in terms of cooling spaces. The calculated lifetime CO_2 emissions for the ASHP, GSHP 1, and GSHP 2 are, respectively, 16, 170, 13,412, and 13,200 kg CO_2.

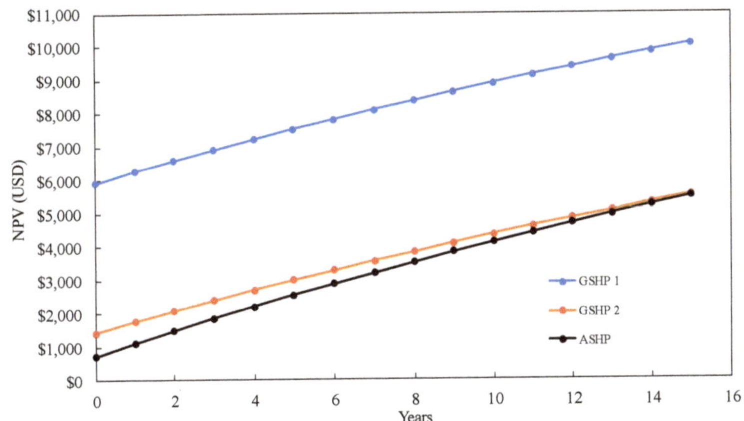

Figure 19. The net present value (NPV) of the heat pumps (in the base case scenario).

Figure 20 presents the calculated NPVs, assuming that an annual power reduction of 40% can be achieved for both GSHPs through the optimization of the systems. In such a case, the annual CO_2 emission is 647.04 kg CO_2/year for each of the GSHPs. It has been observed that the NPV periods for GSHP 2 reach the same value as that for the ASHP after 6.3 years of operation. At the end of its lifespan, GSHP 2 has a 14% cost reduction. The lifetime CO_2 emission for each GSHP is 9705 kg CO_2. However, it is noted that in both scenarios, the cost of GSHP 1 is too high, owing to the high system cost, making the GSHP 1 system unfavorable from an economic point of view, unless the heat pump cost is significantly lowered. Beside the aforementioned cost analyses, to evaluate the actual long-term sustainability of GSHP application, life cycle assessment can be adopted to assess the environmental effects of the GSHP [15,70–72]. Furthermore, long-term data (i.e., year-round experimental data) are required, such that an annual evaluation and comparison of heat pump systems can be carried out.

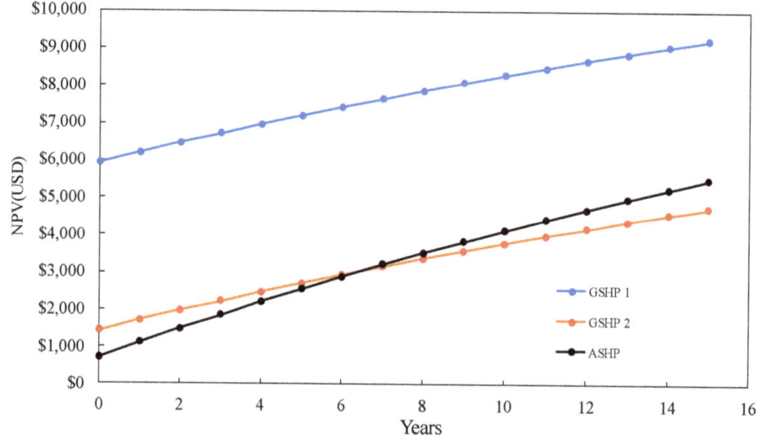

Figure 20. NPV of the heat pumps assuming the GSHP uses 40% less power than the ASHP.

Comparing both GSHPs, it is important to note that the major cost difference relies on the heat pump cost, since both systems used similar ground heat exchangers. The ratio of the heat exchanger and installation cost to the heat pump cost were 10% and 60% for GSHP 1 and GSHP 2, respectively. That is to say, GSHP 1 costs six times more than GSHP 2. Thus, it is necessary to lower the heat pump cost in order to make GSHP application in southeast Asian countries become economically attractive. Cost reduction can be achieved in several ways, such as local manufacturing of GSHP systems, including the heat exchangers, and considering several factors, such as those outlined below:

- Thailand and other southeast Asian countries are experiencing rapid economic and industrial growth [73].
- The ground-source heat pump is a relatively mature technology.
- A single-function heat pump (only for cooling) is less complicated than those with reversible functions (cooling and heating).
- Labor cost is relatively low; installation cost can be further reduced by a proper arrangement and design.

5.4. Future Study

Despite the present study providing early insight into the application of GSHPs in a hot and humid climate, critical analyses are not possible owing to the limited data. Improvements, mainly concerning the data acquisition, thus need to be considered in the future.

Long-term analyses must be conducted for a full year, to clarify the effects of climatic variation on the performance of a heat pump. It is interesting that recent research on GSHP applications in

Bangkok, using the vertical borehole exchanger, showed that an electricity reduction of 30% or more can be achieved [39].

All the aforementioned considerations support the optimization of GSHP in terms of the compressor and heat exchanger capacity, GHE length, size, and separation, while taking into account the building cooling load characteristics and ground conditions.

6. Conclusions

This study presented the application of GSHP in a tropical country (i.e., Thailand) using shallow horizontal heat exchangers. The performances of the two GSHPs were compared to the ASHP in a two-month experiment. The results demonstrate that during the hot season in Thailand, GSHPs with shallow heat exchangers perform better than ASHPs. GSHP 1 and 2 consumed 17.1% and 18.4% less electricity, respectively, than ASHPs. Likewise, the CO_2 emissions could be reduced at a similar rate. During the GSHPs' operational period, the ground temperatures were higher compared to the average daily air temperatures. A significant ground temperature increment was observed at the position near to the heat exchanger, as a result of heat rejection during the GSHPs' operation. Intermittent use of a GSHP and an ASHP provide enough time for the ground to dissipate heat. In addition, cost evaluation revealed that, considering a lifetime of 15 years, investment in GSHPs had no economic advantage over investment in ASHPs. The realization of economic benefit requires a reduction of capital cost. One alternative to achieve this goal is by local manufacturing of the GSHPs.

Further study will focus on long-term performance while upgrading data acquisition, so that other crucial data can be obtained for advance analyses.

Author Contributions: All of the authors contributed to publishing this paper: conceptualization, I.T., S.C. (Srilert Chotpantarat), Y.U., K.Y., and P.C.; methodology, I.T. and A.W.,; validation, A.W., S.C. (Sasimook Chokchai), and I.T.; formal analysis, A.W., S.C. (Sasimook Chokchai), and I.T.; data curation, S.C. (Sasimook Chokchai) and I.T.; writing—original draft preparation, A.W.; writing—review and editing, A.W.; visualization, A.W. and S.C. (Sasimook Chokchai); supervision, Y.U., K.Y., P.C., and S.C. (Srilert Chotpantarat).

Funding: This research is financially supported by (i) The Leading Initiative for Excellent Young Researcher (LEADER), Ministry of Education, Culture, Sport, Science and Technology, Japan; (ii) the National Research University Project, Office of Higher Education Commission (WCU-58-017-EN and NRU59-052-EN); and (iii) the Thammasat University Research Fund under the Research University Network (RUN) initiative.

Acknowledgments: The authors would like to thank to Chulalongkorn University for providing research facilities. Also, Narongsuk Sodsaard of the Faculty of Science, Chulalongkorn University, and Podchana Jamngoen of the Department of Geology, Chulalongkorn University, for technical support on GSHP modification and non-technical assistance at the Saraburi site, respectively.

References

1.   IEA. *Southeast Asia Energy Outlook 2017*; IEA: Paris, France, 2017.

2.   Ministry of Energy of The Kingdom of Thailand. *Energy Statistic of Thailand*; Ministry of Energy of The Kingdom of Thailand: Bangkok, Thailand, 2018.

3.   Lee, Z.H.; Sethupathi, S.; Lee, K.T.; Bhatia, S.; Mohamed, A.R. An Overview on Global Warming in Southeast Asia: CO2emission Status, Efforts Done, and Barriers. *Renew. Sustain. Energy Rev.* **2013**, *28*, 71–81. [CrossRef]

4.   World Air Conditioner Demand by Region. Available online: https://www.google.com.tw/url?sa=t[&rct=j&q=&esrc=s&source=web&cd=1&cad=rja&uact=8&ved=2ahUKEwi5xO3GrLDhAhXUc94KHWfCCesQFjAAegQIAxAC&url=https%3A%2F%2Fwww.jraia.or.jp%2Fenglish%2FWorld_AC_Demand.pdf&usg=AOvVaw1q-QotBU5B0pLSdNOuxDBq (accessed on 24 February 2019).

5.   Kwong, Q.J.; Adam, N.M.; Sahari, B.B. Thermal Comfort Assessment and Potential for Energy Efficiency Enhancement in Modern Tropical Buildings: A Review. *Energy Build.* **2014**, *68*, 547–557. [CrossRef]

6.   Kojima, M.; Watanabe, M. *Effectiveness of Promoting Energy Efficiency in Thailand: The Case of Air Conditioners*; IDE Discussion Papers No. 577; Institute of Developing Economies: Chiba, Japan, 2016.

7. Cazelles, P. *Promotion of Higher Efficiency Air Conditioners in Asean: A Regional Policy Roadmap*; European Copper Institute: Brussels, Belgium, 2015.

8. Sahakian, M. *Keeping Cool in Southeast Asia Energy Consumption and Urban Air-Conditioning*; Palgrave Macmillan: Basingstoke, UK, 2014.

9. Supasa, T.; Hsiau, S.-S.; Lin, S.-M.; Wongsapai, W.; Wu, J.-C. Household Energy Consumption Behaviour for Different Demographic Regions in Thailand from 2000 to 2010. *Sustainability* **2017**, *9*, 2328. [CrossRef]

10. Sriamonkul, W.; Intarajinda, R.; Tongsuk, N.; Saengsuwan, S. Life Cycle Cost Analysis of Air Conditioning System for Residential Sector in Thailand. *GMSARN Int. J.* **2011**, *5*, 131–138.

11. Parkpoom, S.; Harrison, G.P. Analyzing the Impact of Climate Change on Future Electricity Demand in Thailand. *IEEE Trans. Power Syst.* **2008**, *23*, 1441–1448. [CrossRef]

12. Arifwidodo, S.D.; Tanaka, T. The Characteristics of Urban Heat Island in Bangkok, Thailand. *Procedia Soc. Behav. Sci.* **2015**, *195*, 423–428. [CrossRef]

13. Arifwidodo, S.; Chandrasiri, O. *Urban Heat Island and Household Energy Consumption in Bangkok, Thailand*; Elsevier B.V.: Amsterdam, The Netherlands, 2015; Volume 79.

14. Genchi, Y. New Countermeasures for Mitigating the Heat Island Effect - District Heat Supply Systems, Underground Thermal Energy Storage, Systems Using Underground as a Heat Sink (in Japanese). *J. Jpn. Soc. Energy Resour.* **2001**, *22*, 306–310.

15. Genchi, Y.; Kikegawa, Y.; Inaba, A. CO2payback-Time Assessment of a Regional-Scale Heating and Cooling System Using a Ground Source Heat-Pump in a High Energy-Consumption Area in Tokyo. *Appl. Energy* **2002**, *71*, 147–160. [CrossRef]

16. Dasare, R.R.; Saha, S.K. Numerical Study of Horizontal Ground Heat Exchanger for High Energy Demand Applications. *Appl. Therm. Eng.* **2015**, *85*, 252–263. [CrossRef]

17. Congedo, P.M.; Colangelo, G.; Starace, G. CFD Simulations of Horizontal Ground Heat Exchangers: A Comparison among Different Configurations. *Appl. Therm. Eng.* **2012**, *33–34*, 24–32. [CrossRef]

18. Mustafa Omer, A. Ground-Source Heat Pumps Systems and Applications. *Renew. Sustain. Energy Rev.* **2008**, *12*, 344–371. [CrossRef]

19. Kim, M.-J.; Lee, S.-R.; Yoon, S.; Jeon, J.-S. Evaluation of Geometric Factors Influencing Thermal Performance of Horizontal Spiral-Coil Ground Heat Exchangers. *Appl. Therm. Eng.* **2018**, *144*, 788–796. [CrossRef]

20. Fujii, H.; Nishi, K.; Komaniwa, Y.; Chou, N. Numerical Modeling of Slinky-Coil Horizontal Ground Heat Exchangers. *Geothermics* **2012**, *41*, 55–62. [CrossRef]

21. Fujii, H.; Yamasaki, S.; Maehara, T.; Ishikami, T.; Chou, N. Numerical Simulation and Sensitivity Study of Double-Layer Slinky-Coil Horizontal Ground Heat Exchangers. *Geothermics* **2013**, *47*, 61–68. [CrossRef]

22. Di Sipio, E.; Bertermann, D. Soil Thermal Behavior in Different Moisture Condition: An Overview of ITER Project from Laboratory to Field Test Monitoring. *Environ. Earth Sci.* **2018**, *77*, 1–15. [CrossRef]

23. Bertermann, D.; Klug, H.; Morper-Busch, L. A Pan-European Planning Basis for Estimating the Very Shallow Geothermal Energy Potentials. *Renew. Energy* **2015**, *75*, 335–347. [CrossRef]

24. Neuberger, P.; Adamovský, R.; Šedová, M. Temperatures and Heat Flows in a Soil Enclosing a Slinky Horizontal Heat Exchanger. *Energies* **2014**, *7*, 972–987. [CrossRef]

25. Demir, H.; Koyun, A.; Temir, G. Heat Transfer of Horizontal Parallel Pipe Ground Heat Exchanger and Experimental Verification. *Appl. Therm. Eng.* **2009**, *29*, 224–233. [CrossRef]

26. Leong, W.; Tarnawski, V.; Aittomäki, A. Effect of Soil Type and Moisture Content on Ground Heat Pump Performance: Effet Du Type et de l'humidité Du Sol Sur La Performance Des Pompes à Chaleur à Capteurs Enterrés. *Int. J. Refrig.* **1998**, *21*, 595–606. [CrossRef]

27. Chua, K.J.; Chou, S.K.; Yang, W.M. Advances in Heat Pump Systems: A Review. *Appl. Energy* **2010**, *87*, 3611–3624. [CrossRef]

28. Lazzarin, R.M.; Castellotti, F. A New Heat Pump Desiccant Dehumidifier for Supermarket Application. *Energy Build.* **2007**, *39*, 59–65. [CrossRef]

29. Aynur, T.N.; Hwang, Y.; Radermacher, R. Integration of Variable Refrigerant Flow and Heat Pump Desiccant Systems for the Cooling Season. *Appl. Therm. Eng.* **2010**, *30*, 917–927. [CrossRef]

30. Ozgener, O.; Hepbasli, A. A Parametrical Study on the Energetic and Exergetic Assessment of a Solar-Assisted Vertical Ground-Source Heat Pump System Used for Heating a Greenhouse. *Build. Environ.* **2007**, *42*, 11–24. [CrossRef]

31. Han, Z.; Zheng, M.; Kong, F.; Wang, F.; Li, Z.; Bai, T. Numerical Simulation of Solar Assisted Ground-Source Heat Pump Heating System with Latent Heat Energy Storage in Severely Cold Area. *Appl. Therm. Eng.* **2008**, *28*, 1427–1436. [CrossRef]

32. Esen, H.; Esen, M.; Ozsolak, O. Modelling and Experimental Performance Analysis of Solar-Assisted Ground Source Heat Pump System. *J. Exp. Theor. Artif. Intell.* **2017**, *29*, 1–17. [CrossRef]

33. Hawlader, M.N.A.; Rahman, S.M.A.; Jahangeer, K.A. Performance of Evaporator-Collector and Air Collector in Solar Assisted Heat Pump Dryer. *Energy Convers. Manag.* **2008**, *49*, 1612–1619. [CrossRef]

34. Bruelisauer, M.; Meggers, F.; Leibundgut, H. Choosing Your Heat Sink for Cooling in Tropical Climates. In Proceedings of the 5th International Building Physics Conference (IBPC), Kyoto, Japan, 28–31 May 2012; pp. 1339–1345.

35. Permchart, W.; Tanatvanit, S. Study on Using the Ground as A Heat Sink for A 12,000-Btu/h Modified Air Conditioner. *World Acad. Sci. Eng. Technol.* **2009**, *3*, 120–123.

36. Yasukawa, K.; Uchida, Y.; Tenma, N.; Taguchi, Y.; Muraoka, H.; Ishii, T.; Suwanlert, J.; Buapeng, S.; Nguyen, T.H. Groundwater Temperature Survey for Geothermal Heat Pump Application in Tropical Asia. *Bull. Geol. Surv. Jpn.* **2009**, *60*, 459–467. [CrossRef]

37. Yasukawa, K.; Takashima, I.; Uchida, Y. Geothermal Heat Pump Application for Space Cooling in Kamphaengphet, Thailand. *Bull. Geol. Surv. Jpn.* **2009**, *60*, 491–501. [CrossRef]

38. Uchida, Y.; Yasukawa, K.; Tenma, N.; Taguchi, Y.; Suwanlert, J.; Buapeng, S. Subsurface Thermal Regime in the Chao-Phraya Plain, Thailand. *Bull. Geol. Surv. Jpn.* **2009**, *60*, 469–489. [CrossRef]

39. Chokchai, S.; Chotpantarat, S.; Takashima, I.; Uchida, Y.; Widiatmojo, A.; Yasukawa, K.; Charusiri, P. A Pilot Study on Geothermal Heat Pump (GHP) Use for Cooling Operations, and on GHP Site Selection in Tropical Regions Based on a Case Study in Thailand. *Energies* **2018**, *11*, 2356. [CrossRef]

40. Garber, D.; Choudhary, R.; Soga, K. Risk Based Lifetime Costs Assessment of a Ground Source Heat Pump (GSHP) System Design: Methodology and Case Study. *Build. Environ.* **2013**, *60*, 66–80. [CrossRef]

41. Noorollahi, Y.; Bigdelou, P.; Pourfayaz, F.; Yousefi, H. Numerical Modeling and Economic Analysis of a Ground Source Heat Pump for Supplying Energy for a Greenhouse in Alborz Province, Iran. *J. Clean. Prod.* **2016**, *131*, 145–154. [CrossRef]

42. Esen, H.; Inalli, M.; Esen, M. A Techno-Economic Comparison of Ground-Coupled and Air-Coupled Heat Pump System for Space Cooling. *Build. Environ.* **2007**, *42*, 1955–1965. [CrossRef]

43. Esen, H.; Inalli, M.; Esen, M. Technoeconomic Appraisal of a Ground Source Heat Pump System for a Heating Season in Eastern Turkey. *Energy Convers. Manag.* **2006**, *47*, 1281–1297. [CrossRef]

44. Go, G.H.; Lee, S.R.; Yoon, S.; Kim, M.J. Optimum Design of Horizontal Ground-Coupled Heat Pump Systems Using Spiral-Coil-Loop Heat Exchangers. *Appl. Energy* **2016**, *162*, 330–345. [CrossRef]

45. Zhu, Y.; Rayegan, R.; Tao, Y. Case Study of Ground-Source Heat Pump Applications in Hot and Humid Climates. *J. Archit. Eng.* **2014**, *21*, 5014006. [CrossRef]

46. Khedari, J.; Sangprajak, A.; Hirunlabh, J. Thailand Climatic Zones. *Renew. Energy* **2002**, *25*, 267–280. [CrossRef]

47. Vasco, D.A.; Muñoz-Mejías, M.; Pino-Sepúlveda, R.; Ortega-Aguilera, R.; García-Herrera, C. Thermal Simulation of a Social Dwelling in Chile: Effect of the Thermal Zone and the Temperature-Dependant Thermophysical Properties of Light Envelope Materials. *Appl. Therm. Eng.* **2017**, *112*, 771–783. [CrossRef]

48. Staffell, I.; Brett, D.; Brandon, N.; Hawkes, A. A Review of Domestic Heat Pumps. *Energy Environ. Sci.* **2012**, *5*, 9291–9306. [CrossRef]

49. Luo, M.; Cao, B.; Ji, W.; Ouyang, Q.; Lin, B.; Zhu, Y. The Underlying Linkage between Personal Control and Thermal Comfort: Psychological or Physical Effects? *Energy Build.* **2016**, *111*, 56–63. [CrossRef]

50. Hwang, R.L.; Cheng, M.J.; Lin, T.P.; Ho, M.C. Thermal Perceptions, General Adaptation Methods and Occupant's Idea about the Trade-off between Thermal Comfort and Energy Saving in Hot-Humid Regions. *Build. Environ.* **2009**, *44*, 1128–1134. [CrossRef]

51. Feriadi, H.; Wong, N.H. Thermal Comfort for Naturally Ventilated Houses in Indonesia. *Energy Build.* **2004**, *36*, 614–626. [CrossRef]

52. Khedari, J.; Yamtraipat, N.; Pratintong, N.; Hirunlabh, J. Thailand Ventilation Comfort Chart. *Energy Build.* **2000**, *32*, 245–249. [CrossRef]

53. Wong, N.H.; Feriadi, H.; Lim, P.Y.; Tham, K.W.; Sekhar, C.; Cheong, K.W. Thermal Comfort Evaluation of Naturally Ventilated Public Housing in Singapore. *Build. Environ.* **2002**, *37*, 1267–1277. [CrossRef]

54. Wong, N.H.; Khoo, S.S. Thermal Comfort in Classrooms in the Tropics. *Energy Build.* **2003**, *35*, 337–351. [CrossRef]

55. Djamila, H.; Chu, C.M.; Kumaresan, S. Field Study of Thermal Comfort in Residential Buildings in the Equatorial Hot-Humid Climate of Malaysia. *Build. Environ.* **2013**, *62*, 133–142. [CrossRef]

56. Nicol, F. Adaptive Thermal Comfort Standards in the Hot-Humid Tropics. *Energy Build.* **2004**, *36*, 628–637. [CrossRef]

57. Bryś, K.; Bryś, T.; Sayegh, M.A.; Ojrzyńska, H. Subsurface Shallow Depth Soil Layers Thermal Potential for Ground Heat Pumps in Poland. *Energy Build.* **2018**, *165*, 64–75. [CrossRef]

58. Chalhoub, M.; Bernier, M.; Coquet, Y.; Philippe, M. A Simple Heat and Moisture Transfer Model to Predict Ground Temperature for Shallow Ground Heat Exchangers. *Renew. Energy* **2017**, *103*, 295–307. [CrossRef]

59. Chalhoub, M.; Philippe, M.; Coquet, Y. A Field Experiment to Assess the Influence of Heat and Mass Transfer at the Soil Surface on Shallow Ground Heat Exchanger Performances. In Proceedings of the World Geothermal Congress, Melbourne, Australia, 19–25 April 2015.

60. Saito, H.; Simunek, J.; Mohanty, B. Numerical Analysis of Coupled Water, Vapor, and Heat Transport in the Vadose Zone. *Vadose Zone J.* **2006**, *5*, 784–800. [CrossRef]

61. Ghasemi-Fare, O.; Basu, P. Influences of Ground Saturation and Thermal Boundary Condition on Energy Harvesting Using Geothermal Piles. *Energy Build.* **2018**, *165*, 340–351. [CrossRef]

62. Gan, G. Dynamic Thermal Performance of Horizontal Ground Source Heat Pumps – The Impact of Coupled Heat and Moisture Transfer. *Energy* **2018**, *152*, 877–887. [CrossRef]

63. Sakai, M.; Toride, N.; Šimůnek, J. Water and Vapor Movement with Condensation and Evaporation in a Sandy Column. *Soil Sci. Soc. Am. J.* **2009**, *73*, 707. [CrossRef]

64. Platts, A.B.; Cameron, D.A.; Ward, J. Improving the Performance of Ground Coupled Heat Exchangers in Unsaturated Soils. *Energy Build.* **2015**, *104*, 323–335. [CrossRef]

65. Agrawal, K.K.; Misra, R.; Yadav, T.; Agrawal, G.D.; Jamuwa, D.K. Experimental Study to Investigate the Effect of Water Impregnation on Thermal Performance of Earth Air Tunnel Heat Exchanger for Summer Cooling in Hot and Arid Climate. *Renew. Energy* **2018**, *120*, 255–265. [CrossRef]

66. Lu, N.; Dong, Y. Closed-Form Equation for Thermal Conductivity of Unsaturated Soils at Room Temperature. *J. Geotech. Geoenviron. Eng.* **2015**, *141*, 04015016. [CrossRef]

67. Wang, Z.; Wang, F.; Ma, Z.; Wang, X.; Wu, X. Research of Heat and Moisture Transfer Influence on the Characteristics of the Ground Heat Pump Exchangers in Unsaturated Soil. *Energy Build.* **2016**, *130*, 140–149. [CrossRef]

68. Zhang, T.; Cai, G.; Liu, S.; Puppala, A.J. Investigation on Thermal Characteristics and Prediction Models of Soils. *Int. J. Heat Mass Transf.* **2017**, *106*, 1074–1086. [CrossRef]

69. Go, G.H.; Lee, S.R.; N.V., N.; Yoon, S. A New Performance Evaluation Algorithm for Horizontal GCHPs (Ground Coupled Heat Pump Systems) That Considers Rainfall Infiltration. *Energy* **2015**, *83*, 766–777. [CrossRef]

70. Zhu, Y.; Tao, Y.; Rayegan, R. A Comparison of Deterministic and Probabilistic Life Cycle Cost Analyses of Ground Source Heat Pump (GSHP) Applications in Hot and Humid Climate. *Energy Build.* **2012**, *55*, 312–321. [CrossRef]

71. Saner, D.; Juraske, R.; Kübert, M.; Blum, P.; Hellweg, S.; Bayer, P. Is It Only CO2 that Matters? A Life Cycle Perspective on Shallow Geothermal Systems. *Renew. Sustain. Energy Rev.* **2010**, *14*, 1798–1813. [CrossRef]

72. Bayer, P.; Saner, D.; Bolay, S.; Rybach, L.; Blum, P. Greenhouse Gas Emission Savings of Ground Source Heat Pump Systems in Europe: A Review. *Renew. Sustain. Energy Rev.* **2012**, *16*, 1256–1267. [CrossRef]

73. Vithayasrichareon, P.; MacGill, I.F. Portfolio Assessments for Future Generation Investment in Newly Industrializing Countries - A Case Study of Thailand. *Energy* **2012**, *44*, 1044–1058. [CrossRef]

Article

Perspectives on Consumer Habits with Domestic Refrigerators and Its Consequences for Energy Consumption: Case of Study in Guanajuato, Mexico

Juan M. Belman-Flores, Diana Pardo-Cely, Francisco Elizalde-Blancas, Armando Gallegos-Muñoz, Vicente Pérez-García * and Miguel A. Gómez-Martínez

Engineering Division, Campus Irapuato-Salamanca, University of Guanajuato, Salamanca C.P. 36885, Mexico; jfbelman@ugto.mx (J.M.B.-F.); dianapardocely@gmail.com (D.P.-C.); franciscoeb@ugto.mx (F.E.-B.); gallegos@ugto.mx (A.G.-M.); gomezma@ugto.mx (M.A.G.-M.)
* Correspondence: v.perez@ugto.mx; Tel.: +52-464-647-9940

Received: 23 January 2019; Accepted: 25 February 2019; Published: 5 March 2019

Abstract: This work presents the main behaviors shown in the habits of consumers of domestic refrigerators, which influences the energy consumption of this appliance. This study is based on a series of surveys answered by 200 consumers from four cities in the State of Guanajuato, Mexico. The questions were arranged with the aim of evaluating the general characteristics and usage habits such as refrigerator age, door opening frequency, damper position, load of food supplies, external and internal cleaning habits, and nearby heat sources, among other things. The randomly interviewed consumers were individuals between 20 and 60 years of age, who were interviewed using handmade surveys by experts in the field of refrigeration. In some cases, photographic evidence was gathered from the consumers' refrigerators to represent the typical usage habits. In general, the results show that better usage habits are necessary from an energy point of view. Most consumers agree with adopting best practices for using their refrigerator.

Keywords: domestic refrigerator; consumer habits; energy consumption; good practices, surveys

---

1. Introduction

From early civilizations to present time, the need for preserving a space with a lower temperature than that from the environment, in relation to the handling and preservation of food, has been, and will continue to be a fundamental part of life. Thus, the refrigerator is one of the appliances most widely made because of its essential part in homes. Nowadays, domestic refrigeration based on vapor compression technology represents one of the major energy consumers in any home. This technology (air conditioning systems, refrigerators, among others, based on vapor compression) represents about 17% of the energy demand around the world [1]. Moreover, in 2009, approximately 80 million domestic refrigerators were produced [2]. Currently, around 1.5 trillion domestic refrigerators are in service worldwide [1]. These numbers are quickly changing due to the increase in consumers and the need for this type of appliances to preserve and to handle food.

In Mexico, the domestic sector demands the 26% of the national energy consumption, according to the Fideicommissum for the Energy Saving. Thirty percent of this corresponds to the energy consumption of the refrigerator, which is the second appliance with the most energy consumption in homes [3]. Moreover, in Mexico, 80% of homes own at least one refrigerator, which represents more than 23 million in service.

For the last few years, significant efforts have been made to improve the energy efficiency in domestic appliances. Some methods like energy consumption labelling consider energy efficiency as a guide for the search of energy improvements [4,5]. Coupled with this, a variety of mechanisms can increase the energy efficiency in a refrigerator and researchers are widely studying them; among the most relevant mechanism to mention include: improvements in the design of the main components (e.g., compressor and heat exchangers), adequate thermal insulation, application of the regulations, and analysis of the thermal behaviors at the compartments, among other things [6].

Without forgetting that currently there are drop-in replacements of Hydrofluorocarbons (HFCs), by low Global Warming Potential (GWP), refrigerants, in the Kyoto Protocol and recently in the Kigali agreement, there is the task to reduce the use of refrigerants with high GWP. Moreover, in the EU Regulation No. 517/2014 in domestic refrigerators and freezers, employing HFCs was forbidden with 150 or more as GWP [7]. In this sense, in the literature there are several papers where the use of Hydrocarbons, HCs, Hydrofluoroolefins (HFOs), and natural refrigerants in vapor compression refrigeration systems are focused on domestic refrigerators and freezers [8–12]. In fact, HCs are the most recommended [13]. However, despite the main problem with these refrigerants being their high flammability, they are now used in new domestic refrigerators in Europe. Nonetheless, due to being classified as A3 refrigerants, HCs are forbidden in Japan and USA.

On the other hand, currently in Mexico, the use of HFCs and Hydroclorofluorocarbons (HCFCs), is not forbidden yet. For this reason, in the development of this paper, the refrigerant that is present in homes is R134a.

In relation to the improvement of domestic refrigerators, it is clear that a reduction in the energy consumption of this appliance can cause a greater competitiveness for the maker. Moreover, it reduces the environmental impact caused by this appliance, and of course, a benefit for the consumers' savings [6]. On the other hand, factors like room temperature and relative humidity considerably affect the energy consumption of a domestic refrigerator [14,15] and they also affect the refrigerant type and charge used in the refrigerator [16]. Nevertheless, other factors like: door opening frequency, position of the damper, quantity of food, and compartments cleaning, among other things, depend on the consumers' habits, which play an important role in the cold chain and the adequate preservation of the food. Therefore, the consumer plays the main role in, and is responsible for the energy and thermal behavior in a refrigerator. In the literature, works that comprise the study of some of these factors with a statistic focus based on surveys are available. For example, Janjic et al. [17] studied the conditions under which the food is exposed inside the refrigerators such as temperature, cleanliness, and storage practices. The authors concluded that around half of the participants in the survey have inadequate food storage practices in their refrigerators. Moreover, the inner temperature of the refrigerator was considerably higher in comparison with the temperature recommended by the manufacturer. Geppert and Stammiger [18] evaluated consumer behavior concerning refrigerator usage and the main characteristics of these appliances. They analyzed room temperature conditions, compartment inner temperature, and the sources of heat near the refrigerator; all aspects that influence the performance of this appliance. Based on the results, the authors described a series of recommendations about energy efficiency, and with this, they concluded that the information specifically directed to consumers on this topic is scarce. Further in the reading, the authors extended their study to experimentally evaluate some of the operational factors that reflect the daily use of refrigerators such as room temperature, damper position, and the thermal load influenced by the food quantity. They concluded that the energy consumption was very sensitive to the room temperature and to a lesser extent, the refrigerator inner temperature and the thermal load [19]. Hasanuzzaman et al. [20] evaluated the usage habits in refrigerators such as the damper position and the thermal load, in addition to the room temperature and its influence on heat transfer, and the refrigerator's energy consumption. The authors confirmed that the room temperature and the thermal load represented a major impact on the heat transfer and the energy consumption. On the other hand, James et al. [21] gathered diverse studies performed globally that analyzed factors such as frequency with which doors were opened, cleanliness, food handling and storage, and the age of the refrigerator. The above was with the aim of analyzing the thermal behavior of the refrigerator and the cleanliness of the food in relation to their impact on consumer health.

According to the literature reviewed, it is necessary to explore in greater depth the role played by the consumer in the use of the refrigerator to analyze the impact on energy consumption and therefore propose a better use of the energy in this type of household appliances. Thus, this study analyzes diverse factors that influence the energy behavior of a domestic refrigerator. This work is based on the implementation of a series of surveys that were conducted specifically among consumers that live in cities of the State of Guanajuato, Mexico. In addition, the analysis and the results of this study contribute to providing more information about the usage habits of consumers and their relationship with food handling.

2. Material and Methods

The results presented in this study are based on a series of surveys conducted among consumers in cities of the State of Guanajuato, Mexico. The economic activities are different for each city in the state. In addition, consumers from different socioeconomic classes were selected for this survey with the aim of having a more representative sample from the central area of the country. Each of the participants were informed about the aim of the study, and with their prior consent, interviewed. At the end of the interview, photographic evidence was gathered from the inner compartments and exterior of the participants' refrigerator. The survey was performed among relatives, friends, and neighbors, all of them adults, and in different homes with different numbers of family members. Some interviews were also conducted among small homemade-cooking businesses, known as "cocinas económicas". These are typical places in Mexico where consumers buy homemade takeout food. The surveyor registered the answers on paper and by hand.

2.1. Data Colletion

With the aim of having results that reflect to a greater extent the behavior of the usage habits of domestic consumers' refrigerators, 200 surveys were conducted with participants from four cities in the State of Guanajuato and with different socioeconomic levels. The questions were prepared with the aim of demonstrating the energy behavior due to the usage habits of the consumers, and they were conducted by four expert investigators in the field of refrigeration. The questionnaire had 20 questions (see Appendix A), some of them with multiple-choice answers to complement more the main topic. Among the topics of interest and according to the objective of this study the participants were asked: about their refrigerator brand, the time the consumers owned the refrigerator (age), number of members in the family and who used the refrigerator, sources of heat near the refrigerator, (e.g., windows, stove, microwave oven, television), load of food in the compartments (very full, full, regular, mostly empty), number of times the refrigerator doors were opened, and inner and outer cleanliness, among other things. Finally, photographs were taken from those refrigerators showing the most inadequate usage habits. Some of them are presented in this study for illustrative purposes.

2.2. Statistical Analysis

The survey was conducted for 2 months, from May to June 2018 (temperate season with maximum temperature and relative humidity conditions of 27 °C and 40%, respectively), and once all the information was gathered, they were analyzed using Excel as a spreadsheet tool. In such manner, the tool facilitated creating graphics that reflected the behaviors of the usage habits of the consumers and allowed the analysis of the statistical significance (*p*-value) for the quantitative questions (How long have you had your refrigerator?; How often do you open the refrigerator per day?; and How often do you clean your refrigerator?), as are shown in the next sections. This was with the objective of capturing trends regarding refrigerator energy consumption. Among the graphics shown in this paper, bar charts are presented to compare different values and categories. Additionally, pie charts were used to represent percentages and categories proportions.

3. Results and Discussion

Knowing and quantifying the usage habits of refrigerators in Mexican homes, especially those in the State of Guanajuato, make it possible to project the use given to this appliance and the possible impact on energy consumption. This section highlights the main results of this study.

3.1. Respondents

The information gathered from the interviewed consumers was divided into: gender, age, and the number of people that commonly used the refrigerator. Two-hundred surveys were performed, and from these, 81% corresponded to women and 19% to men. These numbers highlight that, in this region of the country, it is women who generally participate in household activities and who provided most of the information gathered in this study. Regarding the age of the interviewed consumers, they are presented in groupings ranging from 20 to more than 60 years old, as is shown in Figure 1. Dividing the participants ages in this manner helps to understand how much the usage habits with refrigerator have changed from one generation to the next. According to Figure 1, it is shown that half of the participants are relatively young consumers, and their age ranks from 20 to 40 years old. Additionally, Figure 2 shows the number of people inhabiting in the surveyed homes, and that commonly use the refrigerator. The data distribution shows that the higher frequency number of inhabitants is four, which is a typical nuclear family in this part of Mexico.

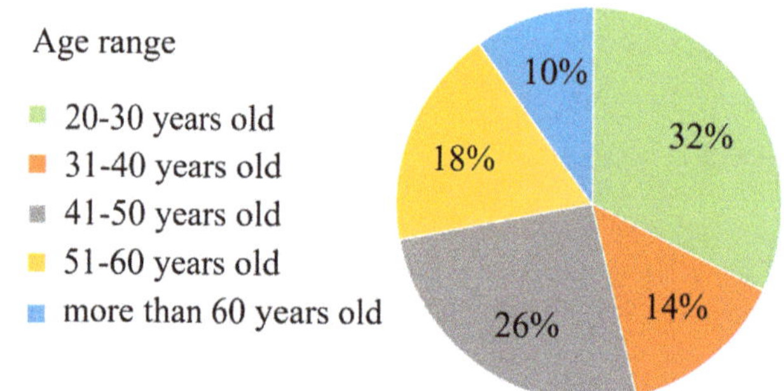

Figure 1. Age range of interviewed consumers.

Figure 2. Number of family members in the surveyed homes.

3.2. Effect of Refrigerator Age

From the 20 questions in the survey, seven are related to the refrigerator's general characteristics, like age and the amount of time the appliance has been in service since the consumer acquired it. The age and the design of the refrigerator play an important role in the energy consumption, as well as in the thermal behavior in the food compartments [17,22]. Modern refrigerators should be more efficient in comparison with older models; they should even be less harmful to the environment, when referring to the type of working fluid [16]. Table 1 shows the age of the refrigerators analyzed for this study

Table 1. Age of the analyzed refrigerators.

	Age Range				
Time (years)	1–5	6–10	11–15	16–20	More than 20
Number of refrigerators	88	62	24	14	12
Percentage of refrigerators (%)	44	31	12	7	6

The table shows that 44% (88 refrigerators) of the consumers had a domestic refrigerator with modern technology and a labeling in accordance with the Mexican regulation NOM-015-ENER-2012 [23]. Additionally, the average age of the refrigerators was presented as eight years, with a confidence interval (CI 95%: 7–9) and a statistical significance $p = 0.026$ ($p <$ 0.05). Studies on the application of energy policies for the replacement of old domestic refrigerators for newer and more efficient ones have shown interesting results for energy savings. It is suggested to replace a refrigerator that has more than 10 years of service. Hence, Table 1 shows that 75% (1–10 years of service) of consumers have adequate refrigerators from an energy efficiency point of view. Nevertheless, the usage habits can considerably affect this energy behavior. In Mexico, a campaign was implemented in 2001 to encourage the consumers to change their refrigerator for a newer and more efficient one. The results of this policy caused energy savings for 4.7 TWh/year, which represents a 33% of the total yearly energy consumption in the domestic sector [24]. Therefore, it was concluded that most of the participants' refrigerators are recent models (refrigerators with less than 10 years of service). The above mentioned is also related to this as a great part of the consumers were relatively young, around 46% of the consumers were between 20 and 40 years old (see Figure 1). Finally, it is worth noticing that a 13% (more than 16 years in service) of the consumers should change their refrigerator for a new one.

3.3. Refrigerator Design

Another general characteristic related to the design of the refrigerator and important for the energy is the defrost system, this is, whether it produces frost or not. The above is simply focused on the comfort of the consumer and the energy consumption that represents a no-frost refrigerator (automatic defrost). According to the results of this study, 80% of the consumers have a no-frost refrigerator at home. This type of refrigerator does not generate frost because it has a thawing device in the freezer, which is generally an electric resistance that through radiation heat transfer removes the frost, which is a permeable coating of ice formed through the inner air humidity. It is worth mentioning that this type of refrigerator consumes more energy, but based on the survey results, consumers prefer them for comfort since it saves them the effort of defrosting it manually. Authors like Bansal et al. [25] estimated that the refrigerators with an automatic defrost system represent an increment in the energy consumption of around 17%. It should be stressed in the present study that all the refrigerators over 16 years in service (see Table 1) present a manual defrosting, and therefore they should theoretically consume less energy. Nonetheless, because of the technology and the implications of their time in service, these refrigerators should be upgraded. According to some studies new refrigerators consume up to 60% less energy than older models for the same size and capacity after eight or more years of use [26].

On the other hand, the design of the magnetic seals in the refrigerator also plays an important role in the energy consumption. They are located on the edges of the refrigerator compartments doors to keep them close. Their usage and wear are caused by the frequency with which the doors are opened, and in many occasions, it is due to the low-quality material with which they are made. If the seals are in good conditions, the refrigerator doors must close hermetically, and they should not present heat gains through them. Seventy percent of the refrigerators reviewed in the survey did not show a visual wear in the seals, and therefore the consumers keep them in good conditions. Studies like the one from Afonso and Castro [27] concluded that an old and damaged seal produces an increase in the energy used by the compressor by 18.5%, when comparing this to a refrigerator with new seals. Thus, the magnetic seals represent a maintenance opportunity that consumers must consider. Thirty percent of the refrigerators reviewed showed damage seals, which could imply a major energy consumption.

Another functional and general characteristic related to the design of the refrigerator is the dissipation of the heat absorbed from the food into the environment during the process of refrigeration. This dissipation is achieved in the condenser, which is a device of major importance in the design of the refrigerator because of its effect in its energy performance. Heat can be dissipated in two ways, the first is through natural convection of the air and the second is through forced convection, which is achieved using a fan. The latter increases the energy consumption in the refrigerator because of the fan. According to survey results in this study, Figure 3 depicts the typical designs of the condensers that the consumers have in their refrigerators. It is observed that 133 consumers have refrigerators in their homes with heat dissipation through natural convection using a wire-on-tube plate condenser. Nevertheless, the wire-on-tube plate condenser is more susceptible to accumulate dirt because of its location, which could affect the heat dissipation, and therefore the energy consumption. Forty-eight refrigerators dissipate the heat using a wire-on-tube coil condenser, this design of heat exchanger requires a ventilation device which slightly increases the energy consumption. Finally, 19 consumers have a refrigerator with a hot-wall condenser at home, which dissipates heat through air natural convection. It is important to mention that most consumers with a hot-wall condenser thought their refrigerator had a problem because of the hot lateral walls on their refrigerator.

3.4. Damper Position

The thermal condition for the storing of food depends on an appropriate distribution of the temperature inside the compartments of the refrigerator, without forgetting the appropriate design of the refrigeration system. Moreover, the thermal behavior is linked to the position of the damper (control element) which controls the air flow and the temperature. Depending on the damper scale, different positions can be selected. For example, some refrigerators are scaled with numbers from 1 to 9 (1 indicates high temperature, 5 medium, and 9 low), and some others are labeled as low, medium, and high. In any of the cases, the position represents a thermal condition for the refrigerator compartments. More sophisticated refrigerators have electronic controls (displays) where the consumer can set the temperature of the compartments. Only 2% of the refrigerators evaluated in the survey present a design of this type. In this study, the position of the damper was observed to predict the internal temperature of the refrigerator. Figure 4 shows these usage habits from the consumers. It was clearly observed that more than half of the consumers (113) keep their refrigerator in a medium position, this probably means that the food compartment keeps a temperature of around 3 °C, while the freezer must have an average temperature of −18 °C; refrigerators with a display are included here since their temperatures were like those mentioned before. These conditions are the typical calibration stablished by manufacturers in Mexico. Twenty-seven consumers adjust their damper in a low position, which represents a high thermal condition in the refrigerator compartments, 7.2 °C in the food compartment and −14.4 °C in the freezer. The rest of the consumers (60) keep their refrigerators in a high position, which means that the refrigerator is in the coldest thermal condition, approximately 1.6 °C in the food compartment and −21 °C in the freezer [16].

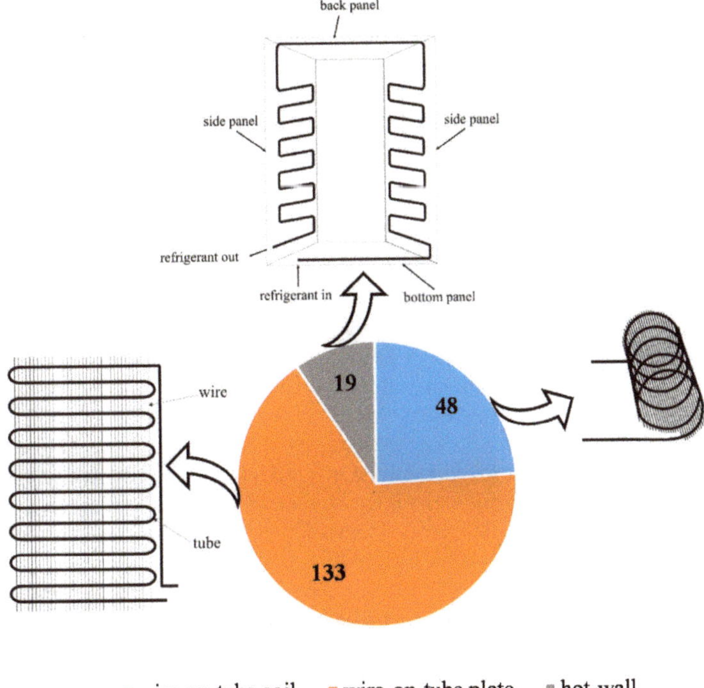

wire-on-tube coil   wire-on-tube plate   hot-wall

Figure 3. Types of condensers in the studied refrigerators.

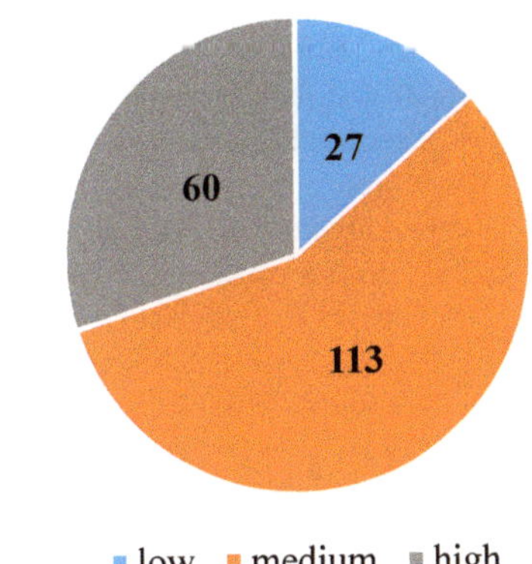

low   medium   high

Figure 4. Damper positions in the analyzed refrigerators.

Most of the consumers do not know the temperature that the damper positions represent, and they prefer to keep the factory set position (position 5 or medium). The thermal conditions in the food compartment based on the damper position for this study are very similar to the ones presented by Geppert and Stamminger [18]. The authors reported that 66.3% of the users adjusted the temperature of the food compartment at 3 °C, while the present study reports that a 56.5% of the consumers keep it at position 5 (3 °C). In contrast, Derens-Bertheau et al. [28] found that the average temperature of food inside the refrigerator was 6.3 °C; one of the possible causes in the thermal difference is the calibration of the appliance and the location of the food inside the compartment. Lower temperatures in the compartments imply a major consumption of energy. For example, Saidur et al. [14] reported that the energy consumption increases around 7.8% per each degree Celsius that the temperature is reduced in the freezer. While Geppert and Stamminger [19] concluded that the energy consumption increases with the decrease of the temperature, from 2 to 5% per each degree Celsius that the temperature is reduced in the food compartment. Similar results were found by Hasanuzzaman et al. [20] with increases of about 46 Wh/day per each degree Celsius that the temperature decreases in the freezer. Hence, according to Figure 4, 30% of the users (60 consumers) are likely to be in a condition of higher energy consumption due to the lower thermal conditions kept in the compartments of the refrigerator.

3.5. Door Opening

As is well known, the main purpose of the refrigerator is related to food preservation. Here the consumer plays an especial role in refrigerator management. The frequency with which the consumer opens the appliance doors to introduce or to withdraw food has an impact on the energy and thermal consumption of the refrigerator. Therefore, the design of the survey considered the number of doors each refrigerator has and the quantity of times these doors were opened. Based on the results from the survey, 35% of the consumers have single-door refrigerators at home. This type of refrigerator usually has the freezer integrated into the food compartment, and because of this, when the refrigerator door is open, the freezer is exposed to an unfavorable thermal condition due to the room temperature which is usually much higher than that of the freezer. The remaining 65% of consumers have two-door refrigerators, which have a physical separation between both the compartments, and they are only opened depending on the necessities of the consumers.

When a refrigerator door is opened, a heat gain is produced due to the heat transfer on the inner surfaces of the refrigerator and the humid air exchange. Furthermore, radiation heat is transferred from the surroundings to the refrigerator. This, along with an elongated time for the opening, causes the compressor to start more frequently to keep the refrigerator inner temperature. In this context, Rahman et al. [29] found a direct relation between the refrigerator temperature and the door opening frequency. Also, Saidur et al. [14] performed an experimental study in which their results showed a considerable influence on the energy consumption due to the door opening frequency, followed by the room temperature from the location of the refrigerator.

Figure 5 depicts the quantity of times that the consumer opened the food compartment door, including the single-door refrigerators. It should be noted that, for the consumer, this question was difficult, as they are not often aware of how frequently they open the refrigerator doors. In view of the above, the consumers were asked to count how often they open the refrigerator doors for a complete day. With these results, it was concluded that 78 consumers open the doors from 21 to 40 times a day. This represents 40% of the survey participants. While 54 consumers (27% of survey participants) open the doors from 41 to 60 times, with an average opening of 43 times a day and an estimate of the confidence interval of 95% (CI: 40–46), representing a level of statistical significance $p = 0.021$ ($p < 0.05$), which indicates that the door opening frequency is mostly concentrated in those ranges for the domestic refrigerator use. In current studies and related to these results, Brennan et al. [30] monitored domestic refrigerators to quantify the door opening frequency. The authors indicated that the most frequent times that doors were opened was 60. Figure 5 shows that nine consumers open doors as often as 100 times in a single day; most of these consumers are those who run a homemade food business, where food circulation is constant in the refrigerator.

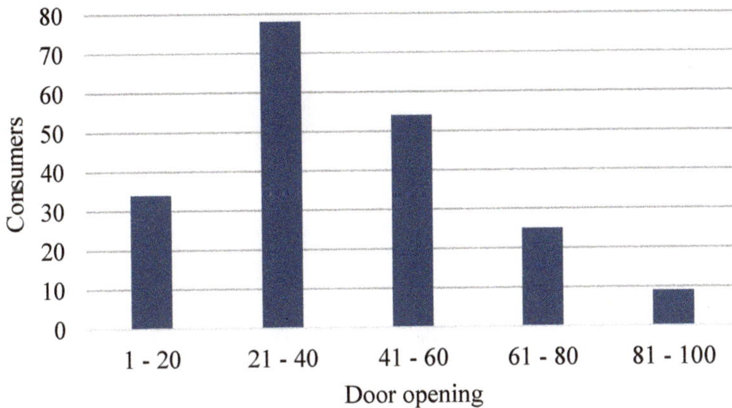

Figure 5. Door opening frequency ranges in the food compartments.

Regarding the times that the freezer door is opened, Figure 6 shows this habit behavior. Note that 152 consumers (76% of total) seldom open the freezer, between 0 to 8 times. It is clear that the food compartment door is open more frequently than the freezer door is, due to the type and amount of food that is stored in those spaces.

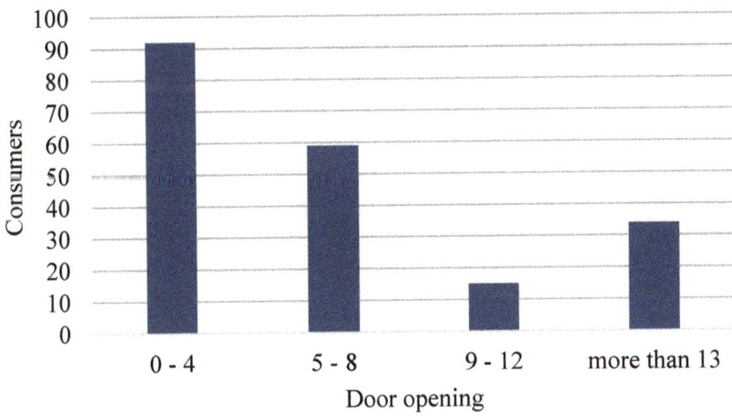

Figure 6. Door opening frequency ranges in the freezer compartment.

The different conditions for the door opening in a domestic refrigerator have been experimentally studied. For example, Khan and Afroz [31] demonstrated that the energy consumption is clearly increased up to 30%, depending on the door opening frequency in comparison with a refrigerator that keeps its door close. The experimental results also show that the ON/OFF cycle of the compressor increased between 2 to 5 times due to the increase in the door opening frequency, which finally caused an increase in the energy consumption of the refrigeration system. Thus, the higher the door opening frequency is, the higher the times the compressor cycle is started; this implies an increase in the energy consumption. Most consumers expressed that they often open the refrigerator door inadvertently, or they simply forget something when circulating the food.

3.6. Presence of Food Load

The correct operation of the refrigerator is linked to an adequate distribution of the temperature and the air flow in the compartments. An appropriate temperature in the refrigerator could extend the food preservation [32], and it could also avoid the development of certain bacteria [33]. It was mentioned before that the thermal behavior is related to the energy consumption of the refrigerator. According to this, the consumer was asked about the practices he follows about the quantity of food stored in the refrigerator. Figure 7 shows the food load conditions in the consumers' refrigerator compartments. The load for the food compartment is indicated in blue and in gray for the freezer.

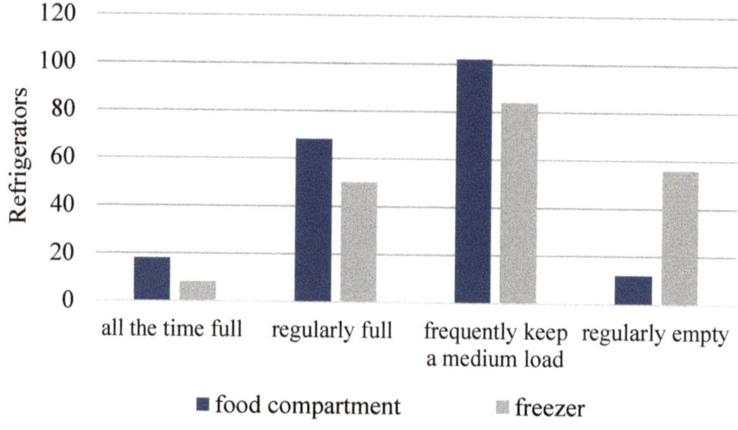

Figure 7. Food load in the refrigerator compartments.

Figure 7 shows that most of the consumers frequently keep a medium load in both compartments of their refrigerator, this represents 51% of the refrigerators evaluated for the food compartment and 42% for the freezer. It could be concluded that this is the most popular use for the refrigerator. A high number of consumers keep their refrigerator regularly full, i.e., they attempt to have a considerable amount of food in both refrigerator compartments. Other consumers keep their refrigerator full all the time; this means that a great quantity of food is always kept in the compartments. Finally, 12 consumers keep the food compartment regularly empty, and 56 keep the freezer empty. In this last habits category, the consumer has scarce perishable food in both compartments. To better illustrate the food load in the refrigerator, Figure 8 shows the aforementioned thermal load conditions. Under this premise is how the analyzed refrigerators were classified for this study.

The quantity of food and its distribution in the compartments could increase their inner temperature due to an inappropriate distribution of the air flow. This also depends on the refrigerator design when referring to the distribution of the air conducts. In the literature, some studies have analyzed the influence of the food load on the energy consumption in the refrigerator. One of the conclusions is that the ON/OFF cycles in the compressor are longer and more frequent in a refrigerator full of food than when it is empty.

All the time full

Regularly full

Frequently keep a medium load

Regularly empty

Figure 8. Thermal load evidence in both the refrigerator compartments.

This happens because the temperature rate of change in the refrigerator is related to the thermal capacity of the food. Hasanuzzaman et al. [20] experimentally evaluated the influence of the food thermal load on the energy consumption, concluding that a significant impact on the energy consumption is observed. Geppert and Stamminger [19] concluded that an additional energy consumption occurs during the food cooling stage. Once the desired temperature is reached, the food load does not considerably affect the energy consumption. It is important to mention that these results may vary depending on the refrigerator design.

The inadequate usage habits of consumers with respect to the quantity of food in the compartments could cause a major energy consumption. According to Figure 8 and based on what was aforementioned, two situations are observed in which a major energy consumption is considered. The first corresponds to the consumers that keep their refrigerator always full, and the second, to those that have their refrigerator regularly empty. A better organization of the food enables a better distribution of the air flow, adequate temperatures, and less energy consumption. Additionally, new compartments designs are being produced with shorter spaces to access the food faster, and without the need to open the complete refrigerator door. This improves and keeps the thermal conditions, which manifest in a lesser energy consumption in the refrigerator.

3.7. Refrigerator Cleaning

It is well known that the refrigerator helps to extend the food preservation, and for this reason, one of the usage habits of the consumers is related to the optimal cleaning conditions. An inappropriate cleaning of the refrigerator compartments frequently causes the food to spoil, and moreover, it enables the development and growth of bacteria and odors [34]. As part of the design of the survey, a question was related to the cleaning frequency in the compartments and on the exterior of the refrigerator. Although the main objective of this study is not the refrigerator cleaning, and its effects in the development of certain bacteria, Figure 9 shows this maintenance habit to extend the information about the consumers' behavior. It was noticed that the consumers kept reasonable cleaning practices, in which the compartments were cleaned every two weeks, monthly, or at least biannually. Only four consumers (2%) cleaned their refrigerator compartments once a year.

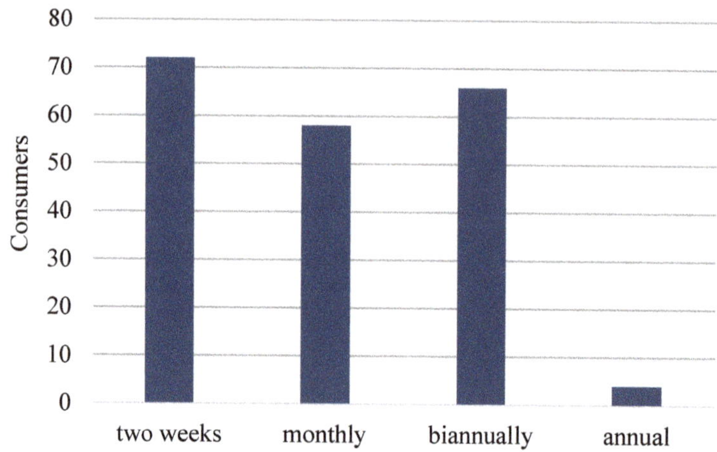

Figure 9. Cleaning frequency in the food compartments.

With respect to the external cleaning, this study highlights the cleaning related to the condenser. Most of the domestic refrigerators analyzed use a wire-on-tube condenser as heat exchanger to dissipate the absorbed heat during the food cooling process (see Figure 3). However, given the fact that a large number of the consumers did not have a cleaning habit for this device, and did not even know its function, the external dirt becomes excessive, mainly composed of dust, fluff, grease, etc. This caused an incorrect operation in all the vapor compression systems, and most refrigerators are based on this technology, since it limits the heat transfer into the environment [35]. For example, Hermes and Melo [36], through the development of a dynamic model, evaluated the effect of the condenser fouling, and they affirmed that this increases the energy consumption by 2%. Belman-Flores [37] experimentally found a variation in the temperature of the refrigerator compartments when comparing a clean condenser with a dirty one. Figure 10 depicts a typical case of fouling found in most of the reviewed refrigerators.

Figure 10. Typical fouling in a condenser.

Figure 11 shows the survey results regarding the condenser cleaning frequency. It was observed that 154 consumers never cleaned the condenser ($p < 0.01$), which causes an inadequate performance in the system, which was noticed in an increase in the energy consumption. Part of the reason for this lack of cleanliness is due to the position of the refrigerator in a particularly enclosed space and the little attention the consumer pays to the back of the appliance.

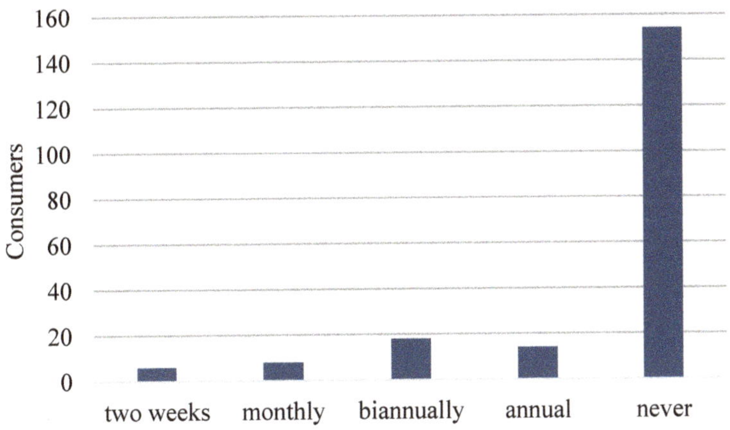

Figure 11. External cleaning of the refrigerator (condenser).

3.8. Refrigerator Close to Heat Sources

Having the refrigerator close to a heat source can affect the temperature in the compartments and increases the energy consumption. This increase is in a range of 0.9 a 1.3% as presented in the work of Lepthien [38]. While conducting the survey, Lepthien paid special attention to the refrigerator conditions and its location with respect to heat sources. For example, 23.5% of the consumers kept their refrigerator away from heat sources. This information is contrary to what other authors have reported; for example, Laguerre et al., [39] reported that 70% of consumers have their refrigerators away from heat sources. While Geppert and Stamminger [18] reported that 34.2% of consumers keep their refrigerator near heat sources. In this study, 76.5% of the consumers have their refrigerator near a heat source. This difference in the results found by other authors could be explained by the architecture of the homes. In Mexico, housing constructions are built in a horizontal manner, and many homes have their kitchen area in front of a gardening area, where it generally has a lot of windows that allow the sunlight to have direct contact with the refrigerator. Moreover, inadequate spacing in kitchens causes consumers to have their refrigerator near heat sources like a stove or a microwave oven. According to the results in this study, Figure 12 shows the distribution of the main heat sources near the refrigerator. The most common heat source near the refrigerator is a microwave oven. The next heat source is when the refrigerator is directly exposed to sunlight through a window, which during several hours of the day represents a continuous heat source. Other consumers have the stove extremely close to the refrigerator, which is a heat source that provides a great quantity of energy to the refrigerator because of its use, whether the burners and/or the oven. Another common habit of the consumers was when they placed a television on top of the refrigerator. Finally, in some cases the consumer had more than one heat source near the refrigerator, which represented 9% of the consumers. Figure 13 depicts evidence of heat sources near the refrigerator. Notice those relevant cases where the consumer had three different nearby heat sources. These usage habits cause an incorrect performance in the refrigerator. Nevertheless, most of them are heat sources used for short periods of time, yet influence the refrigerator performance.

3.9. Preferences in the Purchase of a Refrigerator

Before acquiring a refrigerator, it is important to understand the needs of the family group. Thus, the consumers were asked about their decision standards when buying a refrigerator. Figure 14 shows the most popular decision standards followed by the consumers. The figure shows that 60 (30%) of consumers made their decision to buy a refrigerator based on the energy savings concept, and the rest for other reasons shown in the figure. Energy saving is an important point to consider, as it is necessary to provide more information to the consumer on the importance of buying appliances with adequate energy efficiencies. In Mexico, energy efficiency awareness programs related to refrigerators have been disregarded. For example, campaigns encouraging the consumer to change their old refrigerator for a new one have completely stopped.

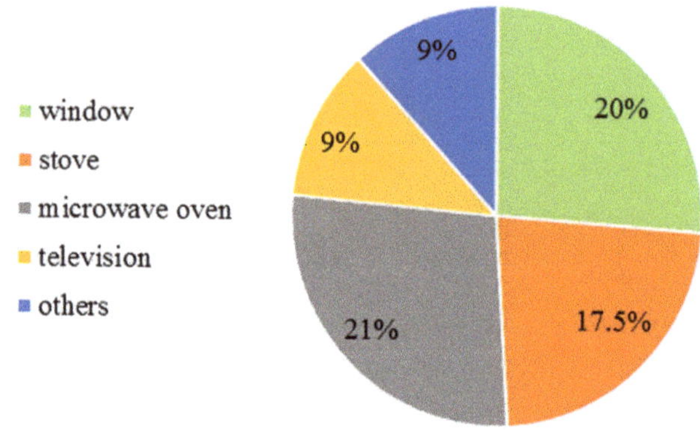

Figure 12. Typical heat sources near the refrigerator.

Figure 13. Typical examples of heat sources near the refrigerator.

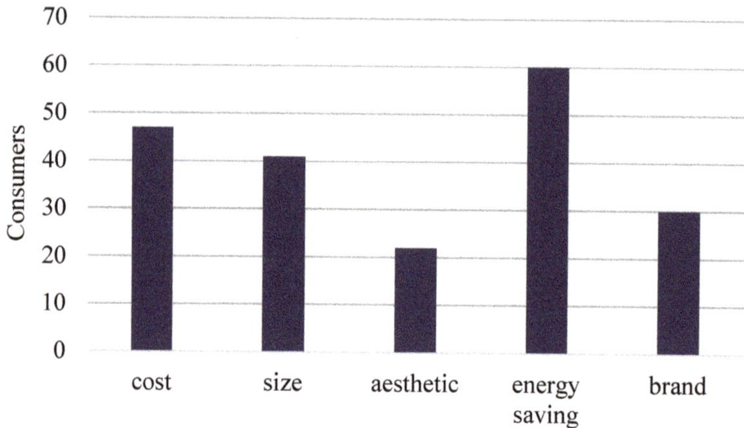

Figure 14. Consumers' decision standards for buying a domestic refrigerator.

Therefore, it was clear that for 60 consumers reducing energy consumption played a major role when deciding to acquire a refrigerator. The refrigerator cost also played a crucial role for the consumer, which frequently depended on the technology which was directly related to energy aspects. It is a little intriguing that size was not an important standard for the consumers when buying a refrigerator, because, based on the survey, around 45% of the consumers claim that their refrigerator was not adequate for their needs. The possible explanation is that the members of the nuclear family extended, or on the contrary, they diminished, and therefore, the size of the refrigerator became inappropriate with respect to their initial needs. This can influence energy consumption due to the aforementioned food load. The brand was another important aspect for the consumer according to the survey. Table 2 shows the refrigerator brand that consumers have at home. It was observed that the refrigerators produced by Mabe were the most popular, followed by Whirlpool, LG, and Daewoo, among other brands. It is important to highlight that Mabe has its main refrigerator factory located in the State of Guanajuato, which encourages a high consumption of domestic appliances in the region and this tendency is like the one reported by other authors on a national level [24].

Table 2. Most popular refrigerator brands in the surveyed homes.

Brand	Refrigerators
Mabe	64
Whirlpool	41
LG	22
Daewoo	16
Samsung	12
Kenmore	11
Frigidaire	10
Koblenz	8
Singer	8
Supermatic	8

3.10. Remarks

The results shown in the previous sections clearly exemplify the main refrigerator usage habits of consumers. It is important to stress that the usage habits considerably influenced the refrigerator energy consumption. According to the results presented in this article, it is concluded that the consumers in this region of Mexico perform an adequate practice in some usage habits. However, for the ones remaining, a great percentage of consumers place their refrigerator near heat sources, which may cause the refrigerator to malfunction. Another inadequate habit is related to the position of the damper, which must be set by the consumer according to the food load or the required thermal condition in the compartments. Another point to highlight is that the consumer must be better informed when buying a refrigerator. The refrigerator must be adequate for the number of family members living at home and the quantity of food they handle. The lack of cleaning of the heat exchanger equipment, like the typical condenser in the evaluated refrigerators, is another of the habits that causes inadequate thermal conditions in the refrigerator compartments, and therefore inadequate energy consumption. Finally, the food load greatly affects the energy consumption in the refrigerator. However, according to the results in this paper, almost half of the consumers maintained their refrigerators with a proper thermal load.

Therefore, it is necessary to diffuse the most adequate usage habits for the refrigerator through the establishment of government or refrigerator manufacturer programs. With this, a sustainable appliance would be achieved/projected through its technology and its use. In this sense, it is expected that manufacturers also commit to the use of refrigerants with low global warming potential [16,40], as it has been pronounced in the Kigali Amendment, which came into effect as of January 2019 [41]. Finally, it is important to mention that the results presented in this paper may be different for consumers from other states in Mexico. Nevertheless, this study is a basis to deepen understanding of the usage habits, and with this, encourage a culture of adequate practices on the usage of refrigerators.

4. Conclusions

This article has presented the behaviors of the main domestic refrigerator usage habits of consumers and their influence on energy consumption. Among the main usage habits discussed in this study, based on a survey, were: the refrigerator age, door opening frequency, damper position, food load, external and internal cleaning, and the nearby sources of heat. The findings indicate that consumers have little knowledge on how refrigerator usage habits affect their appliance energy consumption. Inadequate behaviors were shown during the handling of the food, and this probably caused an increase in energy consumption. This type of study should provide recommendations through manufacturers, which should provide guidance to consumers on better usage practices for this appliance, either through user manuals or signage for use outside or inside the refrigerator. Additionally, dissemination actions could be implemented on energy savings in homes through radio or television campaigns by the federal government. Another possibility would be locally through social service by university students who campaign in homes about the habits surrounding the use and their effects on energy consumption. Most consumers agreed in receiving this kind of orientation and adopting proposals to upgrade their old refrigerators.

Finally, when attempting to achieve a decrease in refrigerator energy consumption and more efficient designs, it is important to know and understand more widely the behavior of the usage habits of consumers. Therefore, presenting the main results of this study is relevant.

Author Contributions: Conceptualization, J.M.B.-F. and D.P.-C.; methodology, F.E.-B. and A.G.-M.; resources, M.A.G.-M.; analysis of data, V.P.-G., J.M.B.-F. and D.P.-C. wrote the paper. All authors have read and approved the final manuscript.

Funding: This research received no external funding.

Acknowledgments: We acknowledge University of Guanajuato for their sponsorship in the realization of this work.

Conflicts of Interest: The authors declare no conflict of interest.

Appendix A

SURVEY (Common usage for the domestic refrigerator)

Refrigerator brand ________________________________

How long have you had your refrigerator? age ____________________

How many doors does the refrigerator have? One _____ Two _____ Three ______

How many family members have access to the refrigerator? ________

Do you know what's the recommended temperature for the refrigerator compartments? No______Yes______ Freezer _________ Freshfood compartment _________

Does your refrigerator generate frost? (do you unplug your refrigerator to defrost it? Yes__ No__

How often do you clean your refrigerator?

Inside: _____ Outside (front, doors): _____ Outside (condenser): _______

Do you unplug your refrigerator to clean it in the inside? Yes_____ No______

Approximately, how often do you open the refrigerator a day?   FF (fresh food compartment)_____ FZ (freezer)_______

Do you ever store hot/warm food in the refrigerator? No ___ Yes ___ (How frequently?) _____

Do you defrost your frozen food in the food compartment? Yes_______ No _____

Heat sources near the refrigerator

Window _____Stove_____ Oven____ Television_____ Others __________________

How full do you keep your refrigerator regularly (thermal load)?

- Food compartment: Very full__ Full__ Regularly full___ Empty most of the time _____

-Freezer: Very full__ Full__ Regularly full___ Empty most of the time _____

Do you consider the refrigerator size adequate for your needs? Yes__ No__

If you bought a refrigerator now, how would you make that decision? According to: Price______ Size______ Looks _____ Energy saving _____ Brand _____

How often do you acquire (buy) a refrigerator? ____________

Surveyor:

Visual wear of the refrigerator seals ___________________________

What's the damper position? Low___ Medium _____ High _____. Number indicated ________ Mechanical or digital _________

What type of condenser is in the refrigerator? Plate___ Espiral___ Hot wall___

Size of the refrigerator: Big ______Medium ______ Small ______

General observations:

References

1. Coulomb, D.; Dupont, J.L.; Pichard, A. The role of refrigeration in the global economy. In *29th Informatory Note on Refrigeration Technologies*; IIR document; International Institute of Refrigeration: Paris, France, 2015.

2. Seong, S. Experimental study of heat transfer characteristics for a refrigerator by using reverse heat loss method. *Int. Commun. Heat Mass Transfer* **2011**, *38*, 572–576.

3. *Balance Nacional de Energía 2012, Secretaría de Energía*; Dirección General de Planeación e Información Energéticas: México, 2013.

4. Vine, E.; Du Pont, P.; Waide, P. Evaluating the impact of appliance efficiency labeling programs and standards: Process, impact, and market transformation evaluations. *Energy* **2001**, *26*, 1041–1059. [CrossRef]

5. Bansal. Developing new test procedures for domestic refrigerators: Harmonization issues and future R&D needs—A review. *Int. J. Refrig.* **2003**, *26*, 735–748.

6. Belman-Flores, J.M.; Barroso-Maldonado, J.M.; Rodríguez-Muñoz, A.P.; Camacho-Vázquez, G. Enhancements in domestic refrigeration, approaching a sustainable refrigerator—A review. *Renew. Sustain. Energy Rev.* **2015**, *51*, 955–968. [CrossRef]

7. The European Parliament and the Council. No. 517/2014 of the European Parliament and the council of 16 april 2014 on fluorinated greenhouse gases and repealing regulation (EC) No. 842/2006 text with EEA relevance. *Off. J. Eur. Union I* **2014**, *150*, 195–230.

8. Aprea, C.; Greco, A.; Maiorino, A.; Masselli, C. The drop-in of HFC134a with HFO1234ze in household refrigerator. *Int. J. Therm. Sci.* **2018**, *127*, 117–125. [CrossRef]

9. Navarro-Esbrí, J.; Mendoza-Miranda, J.M.; Mota-Babiloni, A.; Barragán-Cervera, A.; Belman-Flores, J.M. Experimental analysis of R1234yf as a drop-in replacement for R134a in a vapor compression system. *Int. J. Refrig.* **2013**, *36*, 870–880.

10. Raghunantha Reddy, D.V.; Bhramar, P.; Govindarajulu, K. Hydrocarbon Refrigerant mixtures as an alternative to R134a in Domestic Refrigeration system: The state-of-the-art review. *Int. J. Sci. Eng. Res.* **2016**, *7*, 87–93.

11. Belman-Flores, J.M.; Rangel-Hernández, V.H.; Usón, S.; Rubio-Maya, C. Energy and exergy analysis of R1234yf as drop-in replacement for R134a in a domestic refrigeration system. *Energy* **2017**, *132*, 116–125. [CrossRef]

12. Kasaeian, A.; Hosseini, S.M.; Sheikhopour, M.; Mahian, O.; Yan, W.; Wongwises, S. Applications of eco-friendly refrigerants and nanorefrigerants: A review. *Renew. Sustain. Energy Rev.* **2018**, *96*, 91–99. [CrossRef]

13. Palm, B. Hydrocarbons as refrigerants in small heat pump and refrigeration systems—A revew. *Int. J. Refrig.* **2008**, *31*, 552–563. [CrossRef]

14. Saidur, R.; Masjuki, H.H.; Choudhury, I.A. Role of ambient temperature, door opening, thermostat setting position and their combined effect on refrigerator-Freezer energy consumption. *Energy Convers. Manag.* **2002**, *43*, 845–854. [CrossRef]

15. Ghadiri, F.; Rasti, M. The effect of selecting proper refrigeration cycle components on optimizing energy consumption of the household refrigerators. *Appl. Therm. Eng.* **2014**, *67*, 335–340. [CrossRef]

16. Belman-Flores, J.M.; Rodríguez-Muñoz, A.P.; Gutiérrez Pérez-Reguera, C.; Mota-Babiloni, A. Experimental study of R1234yf as a drop-In replacement for R134a in a domestic refrigerator. *Int. J. Refrig.* **2017**, *81*, 1–11. [CrossRef]

17. Janjic, J.; Katic, V.; Ivanovic, J.; Boskovic, M.; Starcevic, M.; Glamoclija, N.; Baltic, M.Z. Temperatures, cleanliness and food storage practices in domestic refrigerators in Serbia, Belgrade. *Int. J. Consum. Stud.* **2016**, *40*, 276–282. [CrossRef]

18. Geppert, J.; Stamminger, R. Do consumers act in a sustainable way using their refrigerator? The influence of consumer real life behavior on the energy consumption of cooling appliances. *Int. J. Consum. Stud.* **2010**, *34*, 219–227. [CrossRef]

19. Geppert, J.; Stamminger, R. Analysis of effecting factors on domestic refrigerator's energy consumption in use. *Energy Convers. Manag.* **2013**, *76*, 794–800. [CrossRef]

20. Hasanuzzaman, M.; Saidur, R.; Masjuki, H.H. Effects of operating variables on heat transfer and energy consumption of a household refrigerator-Freezer during closed door operation. *Energy* **2009**, *34*, 196–198. [CrossRef]

21. James, C.; Onarinde, B.A.; James, S.J. The use and performance of household refrigerators: A review. *Compr. Rev. Food Sci. Food Saf.* **2017**, *19*, 160–174. [CrossRef]

22. Hassan, H.F.; Dimassi, H.; El Amin, R. Survey and analysis of internal temperatures of Lebanese domestic refrigerators. *Int. J. Refrig.* **2015**, *50*, 165–171. [CrossRef]

23. Secretaría de Energía. *Comisión Nacional para el Uso Eficiente de la Energía. Norma Oficial Mexicana NOM-015-ENER-2012, Eficiencia Energética de Refrigeradores y Congeladores Electrodomésticos. Límites, Métodos de Prueba y Etiquetado*; Secretaría de Energía: Mexico City, Mexico, 2012. (In Spanish)

24. Arroyo-Cabañas, F.G.; Aguillón-Martínez, J.E.; Ambríz-García, J.J.; Canizal, G. Electric energy saving potential by substitution of domestic refrigerators in Mexico. *Energy Policy* **2009**, *37*, 4737–4742. [CrossRef]

25. Bansal, P.; Fothergill, D.; Fernandes, R. Thermal analysis of the defrost cycle in a domestic freezer. *Int. J. Refrig.* **2010**, *33*, 589–599. [CrossRef]

26. General Commission for the Efficient Use of Energy, Secretariat of Energy. 2014. Available online: https://www.gob.mx/conuee#3548 (accessed on 18 July 2018).

27. Afonso, C.; Castro, M. Air infiltration in domestic refrigerators: The influence of the magnetic seals conservation. *Int. J. Refrig.* **2010**, *33*, 856–867. [CrossRef]

28. Derens-Bertheau, E.; Osswald, V.; Laguerre, O.; Alvarez, G. Cold chain of filled food in France. *Int. J. Refrig.* **2015**, *52*, 161–167. [CrossRef]

29. Rahman, S.; Sidik, N.M.; Hassan, M.H.; Rom, T.M.; Jauhari, I. Temperature performance and usage conditions of domestic refrigerator-Freezers in Malaysia. *Hong Kong Inst. Eng. Trans.* **2005**, *12*, 30–35. [CrossRef]

30. Brennan, M.; Kuznesof, S.; Kendall, H.; Olivier, P.; Ladha, C. *Activity Recognition and Temperature Monitoring (ART) Feasibility Study*; Unit Report 25, Social Science Research Unit; Food Standards Agency: London, UK, 2013. Available online: https://www.food.gov.uk/science/research/ssres/foodsafetyss/fs244026a (accessed on 18 June 2018).

31. Khan, M.I.H.; Afroz, H.M. An experimental investigation of door opening effect on household refrigerator; the perspective in Bangladesh. *Asian J. Appl. Sci.* **2014**, *7*, 79–87.

32. Brown, T.; Hipps, N.A.; Easteal, S.; Parry, A.; Evans, J.A. Reducing domestic food waste by lowering home refrigerator temperatures. *Int. J. Refrig.* **2014**, *40*, 246–253. [CrossRef]

33. Garrido, V.; García-Jalón, I.; Vitlas, A.I. Temperature distribution in Spanish refrigerators and its effect on Listeria monocytogenes growth in sliced ready-to-eat ham. *Food Control* **2010**, *21*, 896–901. [CrossRef]

34. Macías-Rodríguez, M.E.; Navarro-Hidalgo, V.; Linares-Morales, J.R.; Olea-Rodríguez, M.A.; Villarruel-López, A.; Castro-Rosas, J.; Gómez-Aldapa, C.A.; Torres-Vitela, M.R. Microbiological safety of domestic refrigerators and the dishcloths used to clean them in Guadalajara, Jalisco, Mexico. *J. Food Prot.* **2013**, *76*, 984–990. [CrossRef] [PubMed]

35. Qureshi, B.A.; Zubair, S.M. The impact of fouling on the condenser of a vapor compression refrigeration system: An experimental observation. *Int. J. Refrig.* **2014**, *38*, 260–266. [CrossRef]

36. Hermes, C.J.L.; Melo, C. Assessment of the energy performance of household refrigerators via dynamic simulation. *Appl. Therm. Eng.* **2009**, *29*, 1153–1165. [CrossRef]

37. Belman-Flores, J.M. Thermal study of the compartments of a domestic refrigerator: Influence of fouling condenser. *Revista Latinoamericana de Innovación Tecnológica Ciencia y Sociedad* **2016**, *1*, 11–14. (In Spanish)

38. Lepthien, K. Umweltschonende Nutzung des Kühlgerätes im Privaten Haushalt. Bachelor's Thesis, Rheinische Friedrich-Wilhelms-Universität Bonn, Institut für Landtechnik, Abteilung Haushaltstechnik, Bonn, Germany, 2000.

39. Laguerre, O.; Derens, E.; Palagos, B. Study of domestic refrigerator temperature and analysis of factors affecting temperature: A French survey. *Int. J. Refrig.* **2002**, *25*, 653–659. [CrossRef]

40. Aprea, C.; Greco, A.; Maiorino, A. HFOs and their binary mixtures with HFC134a working as drop-in refrigerant in a household refrigerator: Energy analysis and environmental impact assessment. *Appl. Therm. Eng.* **2018**, *141*, 226–233. [CrossRef]

41. UNEP. The Kigali Amendment to the Montreal Protocol: HFC Phase-down, Ozon Action Fact Sheet. In Proceedings of the 28th Meeting Parties to Montreal Protocol, Kigali, Rwanda, 10–14 October 2016; pp. 1–7.

MDPI

St. Alban-Anlage 66

4052 Basel

Switzerland

Tel. +41 61 683 77 34

Fax +41 61 302 89 18

www.mdpi.com

Energies Editorial Office

E-mail: energies@mdpi.com

www.mdpi.com/journal/energies